An Analysis of Alternative Approaches to Measuring Multinational Interoperability

Early Development of the Army Interoperability Measurement System (AIMS)

BRYAN W. HALLMARK, CHRISTOPHER G. PERNIN, ANDREA M. ABLER,
RYAN HABERMAN, SALE LILLY, SAMANTHA McBIRNEY,
ANGELA O'MAHONY, ERIK E. MUELLER

Prepared for the United States Army
Approved for public release; distribution unlimited

For more information on this publication, visit **www.rand.org/t/RRA617-1**.

Published by the RAND Corporation, Santa Monica, Calif.

Library of Congress Cataloging-in-Publication Data is available for this publication.

ISBN: 978-1-9774-0743-6

About This Report

This report documents research and analysis conducted as part of the project *Incorporating Interoperability Requirements into Our Exercise Program with Allies and Partners*, sponsored by the Office of the Deputy Chief of Staff, G-3/5/7, U.S. Army. The project aimed to develop a framework for the planning and execution of training and assessments in support of meeting U.S. Army multinational interoperability objectives.

RAND Arroyo Center

This research was conducted within RAND Arroyo Center's Personnel, Training, and Health Program. RAND Arroyo Center, part of the RAND Corporation, is a federally funded research and development center (FFRDC) sponsored by the United States Army.

RAND operates under a "Federal-Wide Assurance" (FWA00003425) and complies with the *Code of Federal Regulations for the Protection of Human Subjects Under United States Law* (45 CFR 46), also known as "the Common Rule," as well as with the implementation guidance set forth in DoD Instruction 3216.02. As applicable, this compliance includes reviews and approvals by RAND's Institutional Review Board (the Human Subjects Protection Committee) and by the U.S. Army. The views of sources utilized in this study are solely their own and do not represent the official policy or position of DoD or the U.S. Government.

Acknowledgments

From our project sponsor's office, we thank LTG Charles Flynn, Deputy Chief of Staff, G-3/5/7, U.S. Army for his ongoing support of this research, LTG(R) Joseph Anderson for his keen interest in measuring interoperability, and MG(R) Christopher P. McPadden for his assistance and support when the project changed direction. We are indebted to our action officer, COL Robert Howieson, Division Chief, Interoperability, Department of the Army's Management Office, Stability and Security Cooperation Division (DAMO-SSC), for his staunch support of the research, along with his invaluable insights, feedback, and guidance. We are grateful to Rich Kurasiewicz, Army G-35-C, Deputy ABCANZ U.S. National Coordinator of ABCANZ Armies and NATO Program. Throughout the research process he and the primary author collaborated on a weekly basis. We greatly appreciate his support, perspectives, and thoughtful insights; it was a pleasure to work with such a professional. We thank Aaron Hill, assessment, monitoring, and evaluation analyst (Headquarters, Department of the Army Deputy Chief of Staff G-3/5/7). From the DAMO-SSC multinational fusion cell (MFC), we thank LTCs Jan Feldman and Claude Boucher for their insights about how to make a measurement valuable to the MFC and in turn the Army.

The remainder of our acknowledgments are in alphabetical order by the title of the respective individuals' organization.

From the ABCANZ Armies Program, we thank COL Andrew Maskell, Chief of Staff, and LTC Simon Boyle, SO1 G7 Training and Lessons. They taught us about the ABCANZ program, provided us copious information to support our analysis of alternatives, and, on several occasions, provided feedback about the Army Interoperability Measurement System (AIMS).

From the Australian Army, we thank Colonel Stuart Cree and Lieutenant Colonel Andrew Keogh for their support during the Talisman Sabre (TS) 19 AIMS pilot. We thank Major Les D'Monte for his assistance with providing AIMS input and a thoughtful review of the AIMS results. We are very grateful to Geoffrey Cooper, Strategy and Capability—Army Lessons, Army Knowledge Centre. Prior to, during, and after TS 19, he was exceptionally supportive. His collaboration and assistance were indispensable to our success during TS 19.

From the Center for Army Analysis, we thank MAJ Chad J. Chapman and Michael V. Pannell. During Joint Warfighting Assessment (JWA) 19, they supported and observed the first AIMS pilot. Their work going through the observation database directly assisted with the creation of records of decisions, and their professional feedback about AIMS was helpful. We are particularly thankful to MAJ Neil Kester. He was the primary architect of Communications Interoperability Capability Appraisal Table (CIRCuIT).

Major Kester's work on CIRCuIT was exceptional, and his contributions were invaluable to the creation of AIMS.

From the Center for Army Lessons Learned, we thank Owen P. McCauley, chief, Partnership and Engagement Division; Phillip Andrews, senior military analyst; and Van Nine, senior military analyst. They provided valuable data, insights, and AIMS feedback during both JWA 19 and TS 19. We appreciate their professionalism and camaraderie.

From Joint Modernization Command, we thank Lieutenant Colonel Kevin Taaffe (United Kingdom, Royal Welsh), G3 Multi-National Interoperability, G3; and CPT Nathan Apticar. Both were immensely helpful leading up to and during JWA 19. In addition, we are especially thankful for the numerous conversations about multidomain operations, interoperability, and AIMS with Lieutenant Colonel Taaffe.

We appreciate the opportunity to collaborate with the Joint Multinational Readiness Center Operations Group. We thank the Commander COL Joseph Hilbert and Deputy Commander LTC Brian McCarthy for their support and insights. We are grateful to CPT Matthew Carstensen, MNI officer, for his assistance and collaboration during AIMS development.

From the Joint Pacific Multinational Readiness Capability, we thank commander of the 196th Infantry Brigade, LTC Jeffrey D. Noll, for his feedback and support, CPT Charles Owen for his assistance, and all of the observer-coach/trainers who provided feedback about and data for AIMS.

From the New Zealand Army, we thank Major Roger Kidd and Warrant Officer Class Two Owen Babb for their insightful comments and review of AIMS output during TS 19.

From the Research and Analysis Center (TRAC), we thank LTC Devin H. Eselius for his assistance prior to and during JWA 19. We also appreciate our discussions with him about measurement and differentiation among assessments, evaluations, and measurement.

From Training and Doctrine Command Capability Manager, Mission Command/Command Posts, we thank Christopher C. Prather, unified action partner branch and U.S. Army Interoperability Proponent, for his perspectives and review of AIMS content.

From the United Kingdom, we thank Major William McCarthy for teaching us how the Multinational Interoperability Assessment Tool

(MIAT) works. During JWA 19 and TS 19, we enjoyed collaborating with Major McCarthy, comparing notes on AIMS and MIAT. We also thank Staff Sergeant Gareth Lyons and Sharon Houston for contributing to AIMS analysis and review during TS 19.

We sincerely thank our two reviewers, Jeff Marquis (RAND) and Mark McDonough, who provided detailed technical reviews that greatly enhanced the report.

Summary

The research reported here was completed in July 2020, followed by security review by the sponsor and the Office of the Chief of Public Affairs, with final sign-off in June 2021.

The 2018 National Defense Strategy (NDS) emphasizes the need for U.S. forces to strengthen alliances, attract new partners, and deepen interoperability with select partners.[1] Being interoperable allows combined forces to produce greater combat power than the sum of their parts by leveraging relative strengths while mitigating relative weaknesses. The Army identified a need to develop an overarching concept for interoperability that includes explicit links between current Army multinational (MN) interoperability (MNI) doctrine and mission command doctrine.[2] With sound concepts, links, and processes in place, the Army can design and implement training exercises that prepare U.S. and MN forces while providing armies an opportunity to observe and measure levels of interoperability. These measures can then be used to determine if countries are achieving their interoperability goals.[3] To support its efforts to create an enduring and integrated interoperability measurement system, the U.S. Army asked RAND Arroyo Center to conduct an analysis of alternatives (AoA) of the measurement system.

There is no widely accepted and standard measurement system; however, several different systems have been developed and used over time to help meet the measurement needs just mentioned. In conjunction with the Deputy Chief of Staff for Operations and Plans, we identified eight different

---

[1] U.S. Department of Defense, *Summary of the National Defense Strategy of the United States of America*, Washington, D.C., 2018.

[2] Army Regulation (AR) 34-1, *Interoperability*, defines interoperability as the "ability to routinely act together coherently, effectively and efficiently to achieve tactical, operational, and strategic objectives" (AR 34-1, *Interoperability*, Washington, D.C.: Headquarters, Department of the Army, April 9, 2020).

[3] In this report, we are focused on developing a system to *measure* interoperability. Translating that measurement into operational outcomes will be done in future studies.

approaches that had been developed or could be modified to measure MNI. We then gathered and analyzed data from a review of materials provided by Army representatives for each approach and information from multiple rounds of interviews with representatives. Although each approach had its advantages, no single one optimally addressed all dimensions. This is why we recommend an approach that combines elements of several alternatives that we analyzed. Table S.1 summarizes the characteristics we propose the new measurement system should possess.

In this report, we discuss dimensions that should be considered when assessing interoperability measurement, delve into the eight alternatives analyzed, propose characteristics that the new system (derived from the dimensions) should possess and how they might be incorporated into the Army's current processes, and detail the Army's evolution of the Army Interoperability Measurement System (AIMS).

We recommended a measurement system that drew on strengths and eliminated weaknesses of other approaches, providing a more enduring and integrated interoperability measurement system. This system would fulfill the Army's need for a standardized and repeatable methodology to identify, evaluate, document, and organize interoperability issues with allies and partners; develop solutions; and communicate and execute those solutions with the Army's senior and operational leaders.

After our AoA, the Army decided to develop a new system for measuring MNI—the AIMS. The Army decided that AIMS would include four distinct components: (1) a quantitative instrument for measuring interoperability levels, (2) a qualitative component to enable capability gap analysis, (3) an automated approach to connect and analyze the data, and (4) exploitation panels. Exploitation panels convene immediately following a training exercise and comprise representatives from all participating countries. These panels collectively work to ensure that the results derived from Component #1 are consistent with "ground truth," and, if there is an issue, they adjudicate the issue. Exploitation panels play a critical role in synthesizing the results from the quantitative and qualitative data analysis to identify and take actions to resolve capability gaps.

The report summarizes the creation of Component #1, pilots of that component conducted at two major training exercises, and how the component's instruments were revised as a result of feedback received during those pilots.

TABLE S.1
Interoperability Measurement System Characteristics Summarized

Characteristic	Recommended Form in the New System
Differentiation from a readiness system	Be a measurement, not an assessment, system.
Ease of use	Be computer- or web-based, enabling observers to easily input data and reducing the amount of subjective judgment required.
Cost incurred	Limit any additional personnel resourcing solely for the purposes of measuring interoperability and leverage extant observer capabilities (including observer, coach, and trainers; collection and analysis teams; and American, British, Canadian, Australian, and New Zealand lessons collection teams).
Consistency with, or similarity to, current Army processes	Measures and data input should look very similar to measures or processes that are already collected during training events.
Output relevant to stakeholders	Have both a quantitative and a qualitative data component with embedded analytic capabilities that automatically calculate interoperability levels by priority focus area,[a] tie levels to the qualitative data, and provide user-defined output to enable capability gap analysis.
Reliability and sustainability	Quantitative measures should be straightforward, directly map to interoperability levels, and be aligned to doctrine to foster universal understanding. To be sustainable, the automated components of the system should be based in a widely available common software framework, such as Excel.
Balance of standardization and flexibility	The system should have both standardized format inputs that allow for comparisons over time and exercise, as well as flexible qualitative inputs that allow for newly emerging challenges to be incorporated.

SOURCE: Interview data and analysis of interviews and AoA.
[a] AR 34-1 defines four priority focus areas for MNI: communication and information systems, which includes knowledge management and information management; intelligence, surveillance and reconnaissance and intelligence fusion; digital fires; and sustainment.

Contents

About This Report iii
Summary vii
Figure and Tables xiii

CHAPTER ONE
Introduction 1

CHAPTER TWO
Considerations 5
Ease of Use 6
Cost Incurred 6
Consistency with, or Similarity to, Current U.S. Army Processes 7
Output Relevance to Stakeholders 7
Balance of Standardization and Flexibility 8
Reliability and Sustainability 9
Differentiation from a Readiness System 10

CHAPTER THREE
Analysis of Alternatives 13
Methodology 15
Training and Evaluation Outlines 17
Center for Army Lessons Learned 19
7th Army Training Command Interoperability Readiness Approach 21
Interoperability Measurement Approach Employed by the Joint Modernization Command at Joint Warfighting Assessment 19 23
ABCANZ Armies' Program Approach 27
Multinational Interoperability Assessment Tool 30
Expert Panel 32

CHAPTER FOUR
Recommendations from the Analysis of Alternatives 35
Conditions to Incorporate in the New System 35
Proposed System Overview 38

CHAPTER FIVE
Early Stages of Army Interoperability Measurement System Development 43
Component #1 Instrument Creation 44
Observer-Based Tool Pilot 1: Joint Warfighting Assessment 19 and Maple Resolve 48
Observer-Based Tool Pilot 2: Talisman Sabre 19 51

CHAPTER SIX
Concluding Remarks 55
Key Characteristics That the System Should Have 55
Army Interoperability Measurement System 56

APPENDIXES
A. Analysis of Alternatives Questions List 59
B. Completed Questions List for All Considered Alternatives 63
C. ART Level I and Level II Tasks Included in AIMS Instruments 93
D. Interoperability in Army Mission Essential Tasks 97
E. Computing Priority Force Area Interoperability Levels 103

Abbreviations 107

References 111

Figure and Tables

Figure

D.1. Example Excerpt of Interoperability Critical Step/Measure..... 98

Tables

S.1. Interoperability Measurement System Characteristics Summarized........ ix
3.1. Levels of Interoperability........ 14
3.2. Example of Center of Excellence Level Input, "ART 3.1: Integrate Fires"........ 26
5.1. ART 3.2 Definitions of Levels of Interoperability........ 45
5.2. Deconstruction of ART 3.2........ 46
5.3. Stem and Root System of ART 3.2........ 47
5.4. Number of Interoperability Measures Piloted at and Revised After Joint Warfighting Assessment 19........ 49
C.1. ART Tasks........ 93
D.1. Mission-Essential Tasks and Supporting Collective Tasks with Multinational Interoperability Considerations........ 99
E.1. Stem/Root System of ART 3.2........ 104

CHAPTER ONE

Introduction

The 2018 National Defense Strategy (NDS) emphasizes the need for U.S. forces to strengthen alliances and attract new partners, which includes deepening interoperability with select partners.[1] To support the NDS, the U.S. Army develops and executes doctrine and guidelines for how its units can achieve interoperability with partners. These include ways to organize to support multinational (MN) operations, how to use mission command processes, and how to inform MN interoperability (MNI) training development and implementation. To achieve interoperability, the U.S. Army conducts and participates in a wide variety of exercises and training activities with allies and partners intended to improve interoperability.

Army Regulation (AR) 34-1, *Interoperability*, defines the term as the "ability to act together coherently, effectively, and efficiently to achieve tactical, operational, and strategic objectives."[2] It further specifies that interoperability activities are an "initiative, forum, agreement, or operation that improves the Army's ability to operate effectively and efficiently as a component of the Joint Force, within an inter-organizational environment, and as a member or leader of an alliance or coalition across the range of military operations."[3] Being interoperable allows coalitions to produce greater combat power than the sum of their parts by leveraging relative strengths while mitigating relative weaknesses. The Army intends for

---

[1] U.S. Department of Defense, *Summary of the National Defense Strategy of the United States of America*, Washington, D.C., 2018.

[2] AR 34-1, *Interoperability*, Washington, D.C.: Headquarters, Department of the Army, April 9, 2020.

[3] AR 34-1, 2020, p. 1.

interoperability to drive fundamental planning and preparation to account for how timing of war has changed; the Army no longer has weeks or months, only mere days, to interoperate with its MN partners in key functions and capabilities. Interoperability includes working with partners, other services, and intergovernmental agencies; however, for purposes of this report, when we use the term, we only mean it in the context of interoperability with MN partners.

The Army identified a pressing need to develop an overarching concept for interoperability that includes explicit links between current Army MNI doctrine and mission command doctrine. Such concepts and links are necessary for the effective integration of interoperability into unit mission planning and MN operations preparation. As of the writing of this report, the Army was developing and disseminating such concepts and processes.

With sound concepts, links, and processes in place, the Army can design and implement training exercises that, first and foremost, train U.S. and MN forces while providing their armies with an opportunity to observe and measure levels of interoperability. These measures would then be used to determine if countries are achieving their interoperability goals. Interoperability goals span four priority focus areas (PFAs): (1) communication and information systems (CIS), which includes knowledge and information management; (2) intelligence, surveillance, and reconnaissance (ISR) and intelligence fusion; (3) digital fires; and (4) sustainment. The PFAs cut across three domains—human, technical, and procedural. Successful interoperability training could take many forms and may include technical integration of systems, building mutual understanding to achieve unity of effort, and/or developing common procedures and understanding at the individual level to operate together coherently. With a process in place that allows for effective design, implementation, and measurement of exercises, the Army and its partners will be able to monitor their progress toward achieving interoperability objectives.

To support its efforts to create an enduring and integrated interoperability measurement system, the Army asked RAND Arroyo Center to conduct an analysis of measurement system alternatives. To date, there has been no widely accepted and standardized measurement system. However, several different systems have been developed and used over time to help meet the measurement needs. In conjunction with the Department of the Army, Military Operations—Stability and Security Cooperation Division (DAMO-SSC), we identified eight approaches that have been developed or could be modified to measure MN interoperability. We then gathered and analyzed data from a review of materials provided by representatives for each approach and multiple rounds of interviews in addition to the materials collected. Although each approach offered advantages, no single approach optimally addressed all dimensions, which is why we subsequently recommend an approach that combines elements of several alternatives analyzed. The recommended measurement system draws on strengths and eliminates weaknesses of other approaches to provide a more enduring and integrated interoperability measurement system. This system fulfills the Army's need for a standardized and repeatable methodology to identify, evaluate, document, and organize interoperability issues with allies and partners; develop solutions; and communicate and execute those solutions with the Army's senior and operational leaders.

The analysis of alternatives (AoA) was one task of a larger six-task project. The other five original tasks did not include any further development or creation of the recommended measurement system. However, soon after the AoA, the Army requested that the ongoing project revise its tasks to directly support the development of its newly named measurement system—the Army Interoperability Measurement System (AIMS). The new tasks executed in 2019 and documented in this report are as follows:

- Develop a preliminary set of measures of performance for the quantitative, observer-based component of the recommended interoperability measurement system.
- Produce a beta version of an automated scoring mechanism.
- Support the AIMS prototype pilot during the Joint Warfighting Assessment (JWA) 19 exercise.[4]

[4] Subsequent to task refinement, the sponsor requested, and we supported, AIMS piloting during two additional training exercises: Maple Resolve 19 in Canada and Talisman Sabre 19 (TS 19) in Australia.

- Refine the quantitative measures and scoring mechanism, and recommend ways to link the quantitative and qualitative data components.

This report documents our AoA, presents the supporting evidence for our measurement system recommendations, and details the early development of AIMS. In Chapter Two, we discuss various dimensions considered when assessing interoperability systems. Chapter Three delves deeper into the eight alternatives analyzed and features a discussion of key limitations of each and takeaways that were incorporated into our recommendations. In Chapter Four, we describe a future interoperability measurement system and how it might be incorporated into the Army's current processes. In Chapter Five, we detail the Army's evolution of AIMS and discuss how it was piloted at JWA 19, Maple Resolve 19, and TS 19; how AIMS tools evolved from one exercise to another; and key takeaways from each training exercise. Chapter Six focuses on next steps and a way forward to successfully implement AIMS. Appendix A provides the list of questions for the analysis of alternatives questions. Appendix B presents a full list of questions and subsequent responses for each assessment. Appendix C provides the ART Level I and Level II tasks included in AIMS instruments. Appendix D covers the MNI conditions statement, excerpt sample critical step/measure, and a list of METs with MNI considerations. And finally, Appendix E outlines the computation of PFA interoperability levels.

CHAPTER TWO

Considerations

In this report, we present a review of options for measuring interoperability across seven primary dimensions: (1) ease of use, (2) cost incurred, (3) consistency with or similarity to current Army processes, (4) output relevance to stakeholders, (5) balance of standardization and flexibility, (6) reliability and sustainability, and (7) differentiation from a readiness system.

Underlying the seven primary dimensions was an initial list of more than 50 subdimensions identified at the beginning of the analysis via discussions with stakeholders, our experiences, and the literature. During our analysis, these subdimensions were grouped according to common themes (e.g., the nature of cost or expensing, primary funder, overhead observer expenses, and travel budget were bundled into the common category of costs incurred). These subdimensions then formed the basis of questions posed to system proponents during the course of site visits, exercise visits, and teleconference discussions. A list of these questions is in Appendix A. The answers to these questions formed the data that enabled us to compare different options according to the Army's organizational interoperability program and policies.[1]

The remainder of this chapter describes each of the primary analysis dimensions.

[1] For example, this is consistent with Execute Order (Headquarters, Department of the Army EXORD) 293-17, which states that commanders should "integrate capabilities and procedures to establish MNI training across Human, Technical, and Procedural Domains required for MN Operations" (U.S. Department of the Army, Execute Order 293-17, *Multinational Interoperability Training*, Washington, D.C., September 2017).

Ease of Use

Ease of use covers both quantitative and qualitative aspects of interoperability. *Quantitative aspects* include the number of rating items present, which helps inform a second qualitative aspect of labor intensity. It may be relevant to consider an interoperability system under multiple scenarios for ease of use, because there is some dynamic interplay between the number of items and the type of data collected, which informs what is ultimately labor-intensive on the part of a rater or team. Another relevant quantitative subcomponent is the collection modality: free-form text, drop-down box, numeric, or stoplight-type chart input. A system could have a large number of items but, depending on the collection modality, remain simple to update or maintain if the methods are simple and unambiguous. Dichotomous response formats (e.g., go or go-go) criteria are an example of a simple collection modality, while free-form narrative text is an example of a more intensive method of measurement.

There is also a qualitative aspect to consider when assessing ease of use and adoption. In this regard, ease was commonly derived from observers' (or data collectors') known use of similar protocols and systems. The more familiar an observer or data collector is with a similar system, the less demanding the schedule was for system familiarization, adoption, and implementation or execution.

Cost Incurred

Interoperability system costs were assessed when possible. Without further defining a funding schedule by exact budgetary allocation or military interdepartmental purchase request, comparisons for costs were made by determining whether a system would be funded under existing arrangements or require new funds. This distinction may not be so clear. For example, a recommended system might be administered at a combatant command–funded exercise, which is seemingly expense-free but may require that the observed unit augment the observer team, at which point costs are incurred. Under such circumstances, net costs are not always immediately knowable. To facilitate the most immediate and useful comparisons across potential options, we look at cost in terms of whether an option is currently funded, the direct unit costs are understood (i.e., some interview discussants were able to identify the dollar or pro rata costs for evaluation), and opportunities exist for measurement options to be conducted in parallel with other rating, readiness, or assessment endeavors.

Consistency with, or Similarity to, Current U.S. Army Processes

Our analysis considers the level of buy-in of Army stakeholders on any prospective interoperability measurement system. This dimension is assessed indirectly by gauging whether an option has features similar to other measurement efforts that the Army is already using (or planning to use). A given interoperability system receives a higher rating in this dimension if the Army is currently using, or has concrete plans to use, similar measures with similar language. In addition, the extent to which similar systems are used more frequently adds to the desirability in this dimension.

Output Relevance to Stakeholders

Stakeholders refers to many U.S. and partner army entities. These include, but are not limited to, units participating in a training exercise and their leaders, the Army Deputy Chief of Staff for Plans and Operations, Army Centers of Excellence, Center for Army Lessons Learned, major commands, and combatant commands. *Relevance* means that the output would readily align with Army strategic and operational interoperability objectives and additional categories of significance to HQDA, such as Army warfighting functions; PFAs; direct correspondence to doctrine, organization, training, materiel, leadership and education, personnel, facilities, and policy (DOTMLPF-P) terminology; or any other comparable system output identified during investigation. If there is additional output relevance, possibly not strictly for Army use but relevant to military partners, then those system outputs are noted as well.

Our analysis of output relevance included output both directly and indirectly produced. We favored output that is directly usable by stakeholders, but, if the output was not directly usable, we considered whether the output measures could be crosswalked to metrics of interest. System outputs can have simple crosswalks to metrics of interest, or there can be multiple intermediate steps required to convert the output into useful metrics. Based on the nature of the intermediate steps, the measurement systems were compared according to the ease of translation to a useful endpoint. Systems having near-automatic conversions that occur within the background of software and scoring systems are preferred over systems that require qualitative, subjective, or human-dependent solutions that exist in addition to the observer's initial output.

Balance of Standardization and Flexibility

The chosen interoperability measurement systems will have to strike a balance between standardization and flexibility. The system should generate standardized and comparable outputs, regardless of the exercise or year. Collection modalities that are quantitative in nature, as opposed to qualitative, are more likely to lend themselves to direct comparisons of one implementation with another.

The Army will also operate with different military partners, with varying numbers of partners in a given operation or exercise, and across operational environments, all of which necessitate a measurement system with immense flexibility to be applied in each setting. Some options may not be suitable in such a variety of venues. For example, an option that can only complete a comprehensive interoperability rating at a formal annual or semiannual exercise may be inflexible in its assessment period because its periodicity is already in lockstep with exercise calendars and budgetary demands established in years prior. Likewise, a system that is strictly implemented outside these major capstone experiences for deployments and unit certifications may possess flexibility but lack consistent processes and, more important, output. Options with ad hoc responses are very flexible, but there is no guarantee that these systems are capable of delivering standardized output. How, exactly, these interoperability options strike the balance between flexibility and standardization, and how these options manage the realities of operational deployment tempos, is addressed in this dimension.

Reliability and Sustainability

Reliability of an interoperability rating system is assessed in several ways—by looking at inter-rater reliability, reliability over different time periods, statistical relevance, or robustness. *Inter-rater reliability*, or the degree to which two observers can see and capture identical impressions, can be enhanced or diminished depending on several other system dimensions discussed here, including output relevance to stakeholders"(dimension 4) and balance of standardization and flexibility (dimension 5). Collection modalities that possess little subjectivity allow for inter-rater reliability to be assessed through direct comparisons of measured values. These collection modalities also lend themselves to greater inter-rater reliability as a whole. However, it is quite difficult, and sometimes not possible, to conduct inter-rater reliability analysis of more-subjective measures (e.g., free-field input).

Reliability over different time periods implies that the interoperability values collected for an event would be identical to the values if the exact same event were replicated at a later time. To illustrate, consider an example of creating a video recording of a training event and having observers complete the measures based on viewing the video. Six months later, the recording is shown to observers, and they record their observations using the same measures. If measured values at both times are identical, then reliability over time would be perfect. This dimension is particularly important if the Army wants to track longitudinal trends with a lengthy amount of time between exercises.

Another potential concern is the length of time that a recorded measure continues to accurately reflect the level of interoperability. In other words, a level measured during a prior exercise may not accurately reflect the "current" level. For example, if two countries participate together in an exercise in 2020 and receive an interoperability level value of *Y*, will the value still be *Y* in 2021 or 2022? The answers to these questions depend both on the measures themselves and on the work the armies do between exercises. Some measures are reasonably straightforward to follow up on—for example, if a measured level reflects that specific equipment is present, then following up to ensure that said equipment is still present is manageable. On the other hand, if the measure reflects availability of personnel or degree of training, the "current" level will be a matter of expert judgment or reassessment at an exercise. One option to consider, in an attempt to maintain transparency about interoperability level sustainment over time, is to have the interoperability system incorporate notices, or disclaimers, about suspension dates for reliability and utility. This is similar to the U.S. intelligence community's habit of noting how quickly information depreciates or is diminished in value before the data or intelligence should be considered unreliable.

Differentiation from a Readiness System

The goal of an Army MNI measurement system is to determine the level of interoperability between the United States and a specific partner (or multiple partners). The system is not intended to be an assessment or performance evaluation system. Although highly nuanced, this is a critical distinction. Interoperability levels are considered necessary, but not sufficient, precursors for readiness. For example, an interoperability level measure could be "Did the United States and MN partners have the appropriate radios with encryption?" whereas a related assessment measure would be "Did the United States and MN partners communicate complete information at the appropriate times?" In this distinction, the interoperability question determines whether the United States and partners had the proper communication systems, while the related assessment question determines whether the right information was communicated at the right times.

Because measuring interoperability is neither an assessment nor an evaluation, interoperability measurement systems should not be used as readiness systems. Interoperability systems can indicate high levels of interoperability for units that may not have high readiness, so it is critical that armies' readiness rating systems are distinctly separate from interoperability levels measurement. Furthermore, we recommend that the Army avoid borrowing readiness terms and structures when creating an enduring interoperability system, thereby avoiding units or MN partners mistaking high or low interoperability levels for equivalent readiness levels.

The chosen interoperability measurement system should maintain this distinction from readiness measurement as both a feature and a well-known disclaimer. If any of the assessed interoperability systems seem to have some ambiguity in this feature, it is noted in this report.

CHAPTER THREE

Analysis of Alternatives

The AoA did not happen in a vacuum and was heavily influenced by events that have transpired over the last several years, during which time interoperability has become an increasing focus within the Army. GEN Mark Milley, then–Chief of Staff of the Army, directed in Army Strategy that interoperability was one of four key lines of effort as part of the endeavor to build partnerships and alliances.[1] In November 2018, the Army was close to finalizing AR 34-1, which set out to define interoperability generally and interoperability levels more specifically. Simultaneously, the Army was working with several key partners to develop interoperability strategic road maps to specify, by partner nation, the desired echelons, interoperability levels, training objectives, and ways to achieve those objectives. In 2019, HQDA was developing and finalizing its Interoperability Campaign Plan, which articulates and synchronizes Army interoperability with unified action partners (UAPs). As of 2018, the Center for Army Analysis (CAA), United Kingdom Ministry of Defence, and the 7th Army Training Command (7ATC), with RAND Arroyo Center, each had independent interoperability measurement efforts running in parallel. Throughout the effort, the Center for Army Lessons Learned (CALL) has remained at the forefront of capturing valuable lessons learned about interoperability and working with Army partners.[2]

---

[1] See Headquarters, Deputy Chief of Staff, G-3/5/7, "The Army Strategy," U.S. Army webpage, October 25, 2018.

[2] An excellent example of these lessons captured is in U.S. Army Combined Arms Center, Center for Army Lessons Learned, *Multinational Interoperability: Reference Guide—Lessons and Best Practices*, Handbook No. 16-18, Fort Leavenworth, Kan., July 2016.

The Army aimed to build an enduring measurement system to measure levels of interoperability consistent with the definitions in AR 34-1. Table 3.1 defines each interoperability level. To standardize interoperability planning and better assess progress toward interoperability goals, the Army recognizes these four levels with UAPs. Level 3 is "integrated" (UAPs are able to fully integrate upon arrival in theater).[3] Level 2 is "compatible" (Army and UAPs are able to interact with each other in the same geographic area in pursuit of a common goal, and they have similar or complementary processes and procedures allowing them to effectively operate with each other). Level 1 is "deconflicted" (Army and UAPs can coexist but do not interact). Level 0 is "not interoperable" (there is no demonstrated interoperability, and partners must operate independently from Army formations and operations).

Given the importance of interoperability and the numerous parallel efforts that were underway, in January 2019, the Army directed RAND Arroyo Center to conduct an AoA as part of a broader project and to present

TABLE 3.1
Levels of Interoperability

Integrated (I-3)	U.S. Army and UAPs are able to integrate upon arrival in theater. Interoperability is network-enabled to provide the full range of military operations capability. UAPs are able to routinely establish networks and operate effectively with or as part of U.S. Army formations.
Compatible (I-2)	U.S. Army and UAPs are able to interact with each other in the same geographic area in pursuit of a common goal. U.S. Army and UAPs have similar or complementary processes and procedures and are able to operate effectively with each other.
Deconflicted (I-1)	U.S. Army and UAPs can coexist but do not interact. Requires alignment of capabilities and procedures to establish operational norms, enabling UAPs and the U.S. Army to complement each other's operations.
Not Interoperable (I-0)	UAPs have no demonstrated interoperability. Command and control (C2) interface with the Army is only at the next higher echelon. UAP formations must operate independently from U.S. Army formations and operations.

SOURCE: AR 34-1, 2020, pp. 2–3.

[3] AR 34-1, 2020.

the findings in early March 2019. This chapter and supporting appendixes document this AoA.

Methodology

Data Collection

The AoA began with the identification of measurement options that were designed or could be adapted to measure interoperability. In collaboration with the sponsor, we created a list of options deemed mature enough to warrant assessment and where the option could be described sufficiently to be assessed. The list included Army training and evaluation outlines (T&EOs) currently used by the Army; CALL collection approaches; an approach developed by RAND Arroyo Center for 7ATC; a data-collection approach employed by Joint Modernization Command (JMC) at JWA 19; CAA's Communications Interoperability Capability Appraisal Table (CIRCuIT); the collection approach developed by the American, British, Canadian, Australian, and New Zealand (ABCANZ) Armies' Program for JWA 19; the UK's Multinational Interoperability Assessment Tool (MIAT); and a generic expert panel approach. We, along with the stakeholders, did not identify any other measurement systems that could be assessed. We then gathered and reviewed all available and relevant information for each approach.[4] We collectively developed a list of questions that were important to determining the origin of the approach, including on the logistics of implementation and data analysis, correlation to interoperability levels, ability to support senior decisionmakers and partners, reliability, and sustainability. A full list of the questions is in Appendix A.

Once this question list was finalized, there were two rounds of telephone interviews conducted with representatives of each agency or approach proponents. Prior to each interview, we answered the questions based on the

[4] All agencies were supportive throughout the entire AoA. They provided documentation in a timely manner, answered interview questions, made themselves available for follow-up questions, and provided feedback. In our judgment, all agencies provided all material that existed.

documentation provided by the agencies. Documentation that was provided to us did not answer all questions we had. Any question not answered by the documentation review was sent to agency representatives in advance of the interviews. These incomplete question lists provided representatives with the questions we would ask during the interview, allowing them to prepare answers ahead of time, if needed.[5] Audio recordings of the interviews were not made; however, many of us took notes during the interviews, and those notes were crosschecked after the interviews.

After each initial interview, we worked to answer all research questions based on the information received, whether by adjusting or adding to answers previously filled in or by filling in answers for questions that were previously blank. Questions still lacking answers were highlighted, and the entire list of research questions and answers were sent to the agency representative. This revised list provided the representatives with an opportunity to (1) confirm or correct the existing answers and (2) prepare for a follow-up interview.

During the second interview, we captured answer changes and collected any additional elaboration that the agency representatives wanted to provide.[6] After this second and final interview, the finalized question list was sent to the proponents to ensure accuracy before including in this report. A full list of questions and subsequent responses for each assessment is included in Appendix B. The remainder of this section will delve into the eight options analyzed, discussing key limitations of each and takeaways that were incorporated into AIMS.

[5] To save time during the first round of interviews, we asked only research questions that were not fully answered from the document review. Each agency was provided an opportunity to review answers to all research questions prior to the second interview.

[6] The answers recorded in our data sheets were fully vetted by the agency representatives. We captured the answers as they were stated in the interview. All changes and modifications provided to us were made.

Training and Evaluation Outlines

Background

T&EOs identify the task, conditions, and standards of performance for collective tasks performed by operational Army units. They are used for training and evaluating unit readiness within the Combined Arms Training Strategy. T&EOs are structured for raters (both in unit and external) to provide dichotomous (go-or-no-go) responses to a number of performance measures that are used as input to determine a proficiency rating (T, T-, P, P-, U) for each mission-essential task (MET) and supporting collective task (SCT) in a unit's mission-essential task list (METL).[7]

Recently, Army EXORD 293-17, *Multinational Interoperability Training*, directed Army commanders to incorporate MNI training objectives into signature training events, including warfighter exercises and Combat Training Center rotations. To inform this directive, the Mission Command Center of Excellence (MCCoE) explored options for including MNI in unit METLs at echelons above brigade. As a result, corps- and theater-level METs were revised to report specifically against interoperability aspects when units conduct MN exercises. It was determined that interoperability itself is not a task; instead, MNI considerations are components of a collective task. Therefore, rather than create separate METs and/or SCTs for interoperability, condition statements and critical performance measures were integrated throughout existing METs and SCTs.[8]

Limitations

Although MNI-specific measures have been added to T&EOs, there are several challenges with using the existing T&EO system to measure interoperability levels. First, the existing MNI performance measures

---

[7] U.S. Department of the Army, *Leader's Guide to Objective Assessment of Training Proficiency*, Washington, D.C., September 2017.

[8] Interviews with MCCoE, February 12, 2019, and March 1, 2019; U.S. Department of the Army, G-3/5/7, *DAMO-TR Briefing: Incorporating Multinational Interoperability Conditions into Corps and Theater Army METs*, Washington, D.C., August 2018, Not available to the general public. See Appendix D for MNI conditions statement, excerpt sample critical step/measure, and a list of METs with MNI considerations.

within the T&EOs are not directly mapped to AR 34-1 interoperability levels or PFAs. Although T&EOs report a given unit's training readiness across all performance measures for specific METs or SCTs, there need to be intermediate steps to get from the T&EO measures to specific MNI PFAs and to subsequently compute the corresponding AR 34-1 interoperability levels for each PFA. Another challenge is that MNI considerations are currently only written for corps- and theater-level METs and SCTs. To measure interoperability at the desired echelons, the relevant METs and SCTs would need to be identified, and MNI measures would need to be written for divisions and brigades.[9]

If the Army were to leverage T&EOs for interoperability measurement without significant adaptation, a significant hurdle would be that the current system of reporting does not allow visibility of a unit's MNI levels. MNI is a condition or consideration during training, so there is no separate MET or SCT to measure MNI specifically. Furthermore, because T&EOs report overall readiness of a MET or SCT, there is no visibility on how a unit performs on individual measures that are highly specific to MNI. For example, the system may report that a unit is partially trained on MET 71-JNT-5100, *Conduct Joint Operations Processes for Joint Task Force*,[10] but there is no systematic way of reporting how the unit performed on specific MNI PFAs embedded in that MET.

Takeaways

Although there are many challenges with using existing T&EOs to measure interoperability, creating measures that are similarly structured yet distinctly separate could be a springboard for designing an interoperability measurement tool. Such an approach would ensure that the tool's measures are doctrinally consistent and would update when T&EO measures are updated to maintain consistency. If aligned well enough with existing

[9] As of this writing, the Army was focused on measuring brigade to theater levels. With AIMS development, there appears to be a move toward including echelons below brigade.

[10] U.S. Army Combined Arms Center, *Conduct Joint Operations Processes for Joint Task Force*, 71-JNT-5100, Fort Leavenworth, Kan., October 2019.

T&EO measures, it could offer units the option of reporting on MNI. Structuring interoperability measures similar to T&EOs may also increase Army-wide acceptance because T&EO language and structure are familiar. Finally, using a similar go-or-no-go rating system lessens the burden on raters and has better inter-rater reliability than a matrix-scoring approach (as is discussed in more detail in the next section). However, one potential consideration is that structuring the measurement system too closely to T&EOs may create the perception that the Army is measuring readiness. As discussed in Chapter Two, it is important that the interoperability measurement system remain distinctly separate from a readiness system and not be perceived as such.

Center for Army Lessons Learned

Background

CALL has a well-developed approach for collecting and disseminating lessons learned and recommendations. Its methodology can be applied at every echelon in training or operational environments where an observer can watch the activity. As described in AR 11-33,[11] *Army Lessons Learned Program*, CALL seeks to improve unit performance through a four-step process of discovery, validation, integration, and assessment of a particular lesson learned. CALL observations are not specifically focused on interoperability, but the team of trained observers that deploys for observation has the capacity to work with both centers of excellence (CoEs) and units to observe and record targeted unit activities, regardless of the topic. Observer teams vary in size depending on the event, from one to 40 personnel; include team augments from various CoEs; and are typically funded through dedicated exercise or training event allotments. CALL observations are submitted to the observed units no more than three months after a training event, are captured in free-form text and written prose, include recommendations, and are written as not-for-attribution for

[11] AR 11-33, *Army Lessons Learned Program*, Washington, D.C.: Headquarters, Department of the Army, June 2018.

the observed units when possible. Established over three decades ago, CALL is well-known within the Army and joint community.[12]

Limitations

The Army Lessons Learned Program generates recommendations that are rooted in the context of training or operational experiences. Although rich with detail and recommended courses of action, the observations are often qualitative and subjective to the observer. Attempts to extract data and generate statistical measures against any CALL content are not readily possible without additional sources of quantitative observations or text algorithms to extract the information. Additionally, comparisons across time and various exercises are not easy to generate because each training event can generate unique sets of observations, the exception being when CALL identifies long-running issues that have not been addressed. Observer teams from CALL are normally dispatched for major Army training events. If CALL is requested outside these scheduled events, there may be some additional costs to units. Spot observations, or accessing an immediate snapshot of interoperability levels, is currently not possible through CALL observations without additional work. Instead, units would need participants in major training events to receive CALL observations in tandem with the exercise and any other assessments.

Takeaways

There are many aspects of the CALL approach that the team considered in designing an interoperability measurement system. CALL can provide context for interoperability levels that is flexible to the needs of the unit, and it is an excellent low- or no-cost option when observation occurs in tandem with Army exercises. CALL observations also scale well across the strategic, operational, and tactical levels of military activity, and, thanks to more than three decades of in-house experience with MN units, CALL possesses the

---

[12] For more information on the evolution of CALL and the Joint Lessons Learned apparatus, see Jon T. Thomas and Douglas L. Schultz, "Lessons About Lessons: Growing the Joint Lessons Learned Program," *Joint Force Quarterly*, Vol. 79, 4th Quarter, October 1, 2015.

ability to generate output across classified and unclassified information domains. CALL observations can also be directly linked to DOTMLPF-P concepts.

CALL's observation, discussion, recommendation, and implication data-capture approach would provide the Army with the types of detail necessary to identify interoperability capability gaps. However, the CALL approach would not yield quantitative interoperability levels. Taking all of this into consideration, we propose that CALL and its current processes be leveraged to provide the qualitative component of an Army interoperability measurement system. CALL Collection and Analysis Teams (CAATs) have, or can readily develop, the PFA expertise to conduct qualitative measurements of interoperability. CAATs already attend many training exercises, making this cost-effective. To link the CAAT qualitative data to quantitative data, some modifications to the Joint Lessons Learned Information System entry form that CALL uses would be required. Using CALL in this capacity would be an efficient use of resident expertise and processes and would contribute to more-rapid adoption of a future measurement system.

7th Army Training Command Interoperability Readiness Approach

Background

In 2018, RAND Arroyo Center developed an interoperability readiness rating approach for the 7ATC, and it was employed during several training rotations at Joint Multinational Readiness Center (JMRC). This system mirrored MET and SCT T&EOs. Approximately two-thirds of the measures came from brigade combat team (BCT) T&EOs, and the remainder were created based on content in CALL handbooks and JMRC lessons learned. Like T&EOs, this approach used go and no-go ratings for interoperability, and the scoring was based on a percentage of measures scored go. Approximately 120 performance measures were identified and fielded in the first version of the instrument provided to the JMRC. Performance measures were organized by warfighting functions, not PFAs. Assessments were made by observer, coach, trainers (OC/Ts). The system was intended to provide an approach analogous to T&EOs that could be used to measure interoperability readiness. The 7ATC model was designed specifically for brigade and units below.

Limitations

This approach has several limitations that prevent it from being optimal for an interoperability measurement system. For one, measures are not directly aligned to AR 34-1 interoperability levels. To generate interoperability levels, this approach computes the percentage of measures that are rated go and then sets cutoffs for each level (zero through three). There is no specific rationale for these cutoffs, and the broader Army enterprise has not weighed in on the best cutoff values; these cutoffs are merely educated guesses. The measures are organized by warfighting functions, not PFAs, although this could be easily changed. A more serious limitation is that there is no structure in place to collect the critical qualitative data necessary to identify DOTMLPF-P capability gaps, which is a capability that many interviewees indicated as critical for a future system. There would have to be another component added to the system to make this connection to capability gaps and DOTMLPF-P solutions.

Takeaways

There are a few key aspects of the 7ATC approach that are highly desirable for an MNI measurement system. For one, it is tailored to MNI, and, although designed for BCTs, the approach is scalable to other echelons. The measures are similar in style and response format to T&EOs, and thus the Army has a level of familiarity with the language and approach used. This approach is straightforward and easy for trained observers to use. There is also little additional cost associated with implementing this method. However, we ultimately do not recommend this approach because (1) the measures were crafted to assess readiness, (2) there is no qualitative data component to support capability gap analyses, and (3) the measures do not readily align to interoperability levels.

Interoperability Measurement Approach Employed by the Joint Modernization Command at Joint Warfighting Assessment 19

Background

During JWA 19, an observation team made up of ABCANZ, JMC, the Research and Analysis Center, CALL, and CoE representatives collected data via a measurement approach designed specifically to measure interoperability during JWA 19. Although the JMC approach was not created for the sole purpose of longitudinal interoperability measurement, we included it in the AoA to identify if it could be adapted for such a purpose. The JMC approach begins with PFAs, and questions are derived from these to address the human, procedural, and technical dimensions. Collection measures are tailored to the event, which would require review and possible revision to ensure that they could be used across exercises. There are five overarching questions with 40 subquestions that vary in response format. Information is collected primarily through observations and surveys, with much of the input being free-form essay-style responses. The data would require significant post-collection processing. Automated metrics, which can include timeliness, lag time, and frequency of common operational picture update, among other items, are also used for some of the data collected, setting JMC apart from many of the other alternatives studied. Analysis is very action-oriented and identifies actionable gaps and associated DOTMLPF-P solutions. The data are not directly linked to interoperability levels other than through an observer's subjective estimate of the appropriate level.

Limitations

Although this approach is valuable in many ways and provided JMC with valuable information for JWA 19, there are some key limitations that prevented us from recommending or adapting it for an enduring system. The data-collection process is very labor-intensive and complex, as is the analysis. The approach was tailor-made for JWA 19 and, although it was effective for that exercise, the methods are not easily transferable to others, especially exercises with fewer observers or with observers who are not

100 percent dedicated to data collection.[13] Observers can select from a large menu of research questions to focus their observations; although this scripting can be valuable to provide focused observations, this approach can also lead to ambiguity. If an observer does not select a research question, should the analyst assume the area represented by the question was a go? Or was it explicitly avoided because it was a no-go? The approach does not have an objective means of identifying interoperability levels. Because of the subjective nature of the measures, they are not statistically robust and would need a complete redesign to be so. The current methodology used is also not sustainable across multiple training events without significant organization.

Takeaways

Although there are some distinct challenges in using the JMC approach in its exact form, the approach has desirable characteristics that have been incorporated into our recommendations. JMC employs a multimethod approach to enable connecting output to solutions and limitations, rendering this approach capable of identifying actionable gaps and associated solutions. The approach is the only system evaluated that has explored the possibility of leveraging automatically generated tactical system measures. It leverages input similar to those of CALL, which includes capture of observation, implications, recommendations, and tags for DOTMLPF-P.

[13] The interoperability observation team at JWA 19 was fully dedicated to data collection, analysis, and writing. This is a very different model than other exercises. In many other exercises, the observers are also coaches and trainers, so data collection is a much smaller part of their duties. An enduring system needs to consider these conflicting labor roles.

Communications Interoperability Capability Appraisal Table

Background

CIRCuIT was developed by the CAA and HQDA DAMO-SSC. It leverages several components to enable an interoperability measurement that connects to the identification of DOTMLPF-P capability gaps. HQDA requested that CoEs use the AR 34-1 definitions for interoperability levels (see Table 3.1) as a template to craft interoperability definitions for specific Army tactical tasks (ARTs).[14] Fifty ARTs were selected and completed, and the output yielded definitions for each of the four interoperability levels for each. An example of the input is shown in Table 3.2.

With CIRCuIT, users would observe a training exercise and, for each ART, assess which level was closest to their observations. In addition to the level, observers would complete a series of web-assessed measures that would enable an analyst to connect the levels to DOTMLPF-P capability gaps. CIRCuIT was designed as a web-based tool that is fairly user-friendly. The levels were meta-tagged to Army warfighting functions (WfFs) and DOTMLPF-P variables, and the ART Level II task outcomes were directly connectable to subjective assessments of limitations and possible solutions.

CIRCuIT shares traits with other options analyzed. Like the CALL option, there would be qualitative data to enable capability gap analysis. Like the T&EO and 7ATC options, CIRCuIT could be directly tied to ARTs, METs, and SCTs. A significant advantage of CIRCuIT over the other options is that the measures explicitly map onto AR 34-1 levels.

Limitations

We found many CIRCuIT characteristics desirable; however, there was a key limitation preventing us from recommending CIRCuIT in its entirety. A great strength of CIRCuIT is that it would yield interoperability levels consistent with AR 34-1. However, because the CIRCuIT definitions reflect AR 34-1 general-level descriptions, they become multifaceted (see Table 3.2).

[14] The ARTs chosen would be commonly exercised in signature training exercises.

TABLE 3.2
Example of Center of Excellence Level Input, "ART 3.1: Integrate Fires"

Integrated (3)	U.S. Army and mission partners conduct targeting within an established working group properly equipped with network-enabled collaboration tools that assist in the synchronizing of target location, identification, classification, tracking, attack guidance, and battle damage assessment. The MPE [mission partner environment] network allows partners to send digital information in near real time that supports the decide, detect, deliver, assess (D3A) targeting process using each mission partner's organic fires C2 systems and support structure. U.S. Army and mission partners are both able to nominate and achieve electronic and computer network attack effects on targets. Little to no liaison officer (LNO) support required.
Compatible (2)	U.S. Army and mission partners actively seek to establish interoperability by leveraging a combination of digital and analog capabilities or methodologies. Networked capabilities are limited to select digital-to-digital touch points often driving human intervention to conduct digital-to-analog-to-digital conversions with some LNO participation. The state of the MPE network requires mission partners to rely on some human dimensions to incorporate procedural controls in order to support the D3A targeting process. Liaison with U.S. network equipment required.
Deconflicted (1)	The mission partner operates with analog systems or leverages outdated, cyber-vulnerable digital capability. The mission partner relies heavily on liaison teams resourced with U.S. equipment and personnel often still requiring analog-to-digital conversions to support the D3A targeting process. The mission partner possesses little institutional knowledge or has limited MPE experience with coalition operations.
Not interoperable (0)	U.S. Army and mission partners do not integrate fires using any digital, analog, or liaison means of communications. The mission partner possesses no institutional knowledge or MPE experience with coalition operations.

SOURCE: Excel file provided by U.S. Army Fires Center of Excellence.

During exercises, observers will see that aspects of several different levels occur simultaneously, and not every aspect for a specific level was present. This requires the observer to make a subjective determination about which level to select because the training observed has elements of multiple levels and no level is fully achieved. Therefore, we anticipate these multifaceted definitions will cause serious challenges for observers and significant reliability issues.

We have seen approaches similar to CIRCuIT used by other parts of the Army. For example, the National Training Center Operations Group has employed the Decisive Action Big 10. This measurement approach poses a macro-level question, such as "conduct direct fire," provides a detailed breakdown of what each score on a scale of 1 to 5 entails, and then has observers score on the 1 to 5 scale based on their individual observations. So, although there is already buy-in within the Army to use a CIRCuIT-like approach, there still remain significant limitations to using this option—namely, the reliability of inter-rater agreement.

A prior unpublished RAND Arroyo Center study compared inter-rater agreement when raters provided responses using different style measures. The first was a measure style that focused on a single element with a dichotomous go-and-no-go format response.[15] The second was a matrix-style approach, such as CIRCuIT. For both response formats, the researchers measured the percentage of agreement between two raters after observing the same training event. They found that the percentage of agreement was greater when using the dichotomous versus the matrix response format. In fact, for the matrix approach, they tested 15 measures, and only three had agreement rates above 50 percent, while the majority of the simpler dichotomous measures had inter-rater agreement above 50 percent.

Takeaways

CIRCuIT has some significant strengths, including direct mapping to AR 34-1 levels; incorporation of qualitative data to support capability gap analysis; a high degree of familiarity within the Army, leading to increased buy-in; and the ability to meta-tag to WfF and DOTMLPF-P variables. With some slight modifications, we recommend that the Army include several of these aspects in its future measurement system.

ABCANZ Armies' Program Approach

Background

The ABCANZ Armies' Program (AAP) exists to improve interoperability in the land domain using human, technological, doctrinal, and procedural

[15] These similar elements are like a single sentence in a level in Table 3.2.

solutions. In support of the end state, the main effort is achieving command interoperability from the battlegroup through to the division with four PFAs—CIS, fires, intelligence fusion, and ISR. For the purpose of interoperability assessment, these four PFAs have been built out to include six, which include a consideration of sustainment and information management and knowledge management (IM/KM).

Critical question lists (CQLs) are used to analyze interoperability levels within a training environment where two or more ABCANZ armies participate. CQLs are focused on the needs of the warfighter and their understanding of interoperability, and the answers to these sets of questions are designed to inform MN training objectives. Specific and more detailed focused question sets (FQSs) are developed by focus area for each MN exercise, further informing capability development within given areas. There is a total of 38 items to answer with a *Y*, *Y(+)*, or *N*. *Y* means that a level of compatible interoperability was achieved; *Y*(+) means that the forces reached a level better than compatible but not quite integrated; and *N* means that units were deconflicted. Although *Y*, *Y*+, and *N* outputs are discrete, in the event that there is *N* response, a narrative is provided explaining why the units were deconflicted. The data-collection process is considered to be very straightforward and easy to implement.

Two methods of data collection are used: direct and indirect. For direct collection, the ABCANZ team implements the assessment with its own assessors. There are also indirect collections that take place, with reports and assessments coming in from the five nations (or portions thereof). In this case, the nations modify their own exercises to accommodate ABCANZ requests and assess interoperability.

Once data are collected, a panel of subject-matter experts (SMEs) comes together to evaluate the information provided, draw conclusions, and come up with major lessons learned. The panel then writes a report detailing how interoperable the nations are. The use of an SME panel helps reduce subjectivity; multiple raters are required to come to a consensus before a report is disseminated, thereby mitigating inter-rater reliability concerns.

Limitations

Although the ABCANZ approach is valuable in many ways, there are some key limitations that prevented us from recommending this method in its entirety. Because it is an ABCANZ tool, the language used is not doctrinal to the U.S. Army, which could lead not only to confusion but also would inevitably result in decreased buy-in from the Army. In addition, using this approach would mean that the Army would not have direct authority over what information is collected when, which is a significant disadvantage when creating a system specifically for the Army's use. Additionally, the system does not measure all four levels of interoperability. By design, it measures whether units are compatible, deconflicted, or not interoperable. It does not measure whether they are integrated. The output is not directly relevant to U.S. stakeholders. Data elements are not linked to U.S. WfF; the human, procedural, and technical aspects of interoperability; ARTs; or DOTMLPF-P. We think measures could be meta-tagged and linked to these attributes with relative ease.

Takeaways

The ABCANZ data-collection and analysis approach is sound, and we recommend that the Army consider including several of its elements in the future measurement system. The ABCANZ approach enables observers to record observations and capture likely limitations and shortfalls, all of which are needed for capability gap analysis. It relies on a straightforward measurement and response format similar to T&EO measures. Being similar to T&EOs is a significant advantage in that the Army is already familiar with this format, so maintaining it would allow for a more seamless transition between assessment methods. Lastly, the ABCANZ approach uses an SME panel to draw conclusions and disseminate lessons learned. Each of these elements should be incorporated into a future measurement system.

Multinational Interoperability Assessment Tool

Background

MIAT is a tool currently in development by the British Army's Future Force Development with an aim to more systematically measure MNI between the British Army and its partners. MIAT does not assess readiness; instead, it measures current levels of interoperability to determine whether they are in line with strategic visions and goals to inform senior leaders' investment decisions. MIAT aims to be an intuitive and objective measurement system that will allow any user (i.e., the rater does not have to be specially trained) from the British Army or its partner nations to rate a unit's interoperability. The tool is being designed for use as a self-assessment, bilateral assessment, or MN assessment, focusing on brigade and echelons above brigade.

MIAT questions are derived from ABCANZ CQLs and FQSs and will use the same three-level response format. The output will be an interoperability rating based on the three ABCANZ interoperability levels: not interoperable, deconflicted, or compatible. Measurement will include observer narratives where appropriate.[16]

In practice, MIAT is a software tool version of the ABCANZ measures that can be loaded onto military-issued laptops. Raters will complete the questionnaire locally, and the observations are then downloaded to a local, portable laptop server and analyzed. This allows for standardized data collection such that each assessment can be compared with output from other assessments. Once developed, the assessment measures will largely remain static over time; however, the system will have the flexibility to add measures when a new capability is added or remove measures that are no longer useful.

Limitations

There are several challenges that could make the MIAT approach difficult to implement as an interoperability measurement system for the Army.

[16] At the time of data collection for this report, and in subsequent observation of MIAT, we observed that the term *integrated* was not being used. As of December 2020, we explained that MIAT does now use integrated.

First, because MIAT is designed to be used by ABCANZ partners, its measures are not necessarily doctrinally consistent with existing Army systems and processes.[17] As a result, as the Army inevitably continues to make changes to U.S. doctrine, it could be difficult to implement parallel changes in an ABCANZ measurement system. The MIAT output will also be based on interoperability levels as defined by ABCANZ rather than AR 34-1. Although this may not be an issue in the long term, in the short term, MIAT does not include an integrated level in its measurement space as AR 34-1 does.[18] MIAT specifies a level for each measure, but, at the time of our AoA, MIAT had not developed a means to rollup measures into a unitary level value for each PFA. That is, *Y*, *Y(+)*, or *N* for each measure is clear; however, if there were ten measures in a single PFA with different levels, there is no articulated aggregation for a single PFA value. In addition, MIAT is designed primarily for brigade units and echelons above brigade. The developers noted that the intent is to have a "stretch downward potential" that would enable the assessment to be used at any desired echelon; however, it appears that extra preparation would be needed to use the assessment at echelons below brigade. These issues are not trivial but could be overcome with some additional effort prior to implementation.

Takeaways

Even amid these challenges, MIAT has several desirable components that the Army should incorporate into its interoperability measurement system. First, if implemented according to its design, MIAT will be exceedingly easy to use, and the cost of use will likely be low. The computer-based system will enable rapid connection of measured levels to qualitative data for subsequent capability gap analysis. The system will be fully standardized, allowing comparison between exercises, and will be flexible enough to add or remove measures when necessary. We recommend

[17] Note that although this is the case, MIAT is not limited to ABCANZ partners and has been used by UK bilateral and multilateral partners and NATO applications.

[18] At the time of data collection for this report, integrated was not being used. This has since changed.

that all of these aspects be considered during the development of an interoperability system.

Expert Panel

Background

Our analysis of expert panels is not based on any particular panel per se but instead on an amalgamation of several examples. Army expert panels include a range of SME-oriented groups that, through a defined scope and timeline, generate recommendations for potential adoption. Expert panels are malleable and, historically, have been summoned when an emergent or episodic issue needs a mix of qualitative and quantitative measures generated to facilitate decisionmaking. A few examples from recent history include the congressionally directed National Commission on the Future of the Army (2016), the Expert Panel on the Future of Army Laboratories (2012), and Military Expert Panel Report on Sea Level Rise and the U.S. Military's Mission (2016).[19] As evidenced by the panel topics, expert panels can either be narrowly defined, as in the case of sea level rise, or exceptionally broad, as in the case of a future army. In either case, a team of experts with diverse backgrounds will investigate research questions, interview stakeholders, and generate a consensus opinion on actions that could be taken to address a problem or task, with substantive cost-benefit analysis evidencing any conclusions.

The nature of the expert panel topic can direct who takes the lead in the panel effort, with more-technical equipment or injury-related phenomena attracting science and technology SMEs and more-operational problems attracting strategy and combat SMEs. Since panel participation is driven in part by the problem, Army expert panels do not reside within any one program of record. Additionally, the top-down (congressionally directed) and bottom-up (Army directed) approach to expert panels suggests that panel

[19] See for example, Ronald Keys, John Castellaw, Robert Parker, Jonathan White, Gerald Galloway, and Christine Parthemore, *Military Expert Panel Report: Sea Level Rise and the U.S. Military's Mission*, Shiloh Fetzek, Caitlin E. Werrell, and Francesco Femia, eds., Washington, D.C.: Center for Climate and Security, September 2016.

leaders, stakeholders, scope, and timeline are all flexible elements in the practice of running a panel.

Limitations

Expert panels are specifically formed to address a specific episodic subject that makes them less desirable as an enduring system. If expert panels were chosen, the design would need to support repeatable interoperability level measurement. Their cost and burden would be greater than the other approaches because expert panels would need to be created, coordinated, managed, and funded on an ongoing basis. Additionally, panels lack an across-time and -exercise objective measurement element. Such elements would need to be developed for a future system.

Takeaways

Expert panels have a tradition within the Army. They lend themselves to the resolution of unique, complex, or persistent problems that the Army faces. Interoperability rating systems, or the broader military interoperability problem set, may have elements that are suited to expert panels when there is an imminent concern or threat that requires a multidisciplinary response. However, standing requirements, such as measuring regular interoperability levels of units across echelon, do not seem immediately suited to expert panel resolution. The Army's history of using expert panels has demonstrated a degree of flexibility, that there is not a one-size-fits-all approach in expert panel usage, and that time, as a resource, is essential to problem-solving (the aforementioned panels ranged from months to years in terms of execution). None of these features readily lends itself to the demands of regular and repeatable interoperability level measurement. Any attempt to apply expert panels in the domain of interoperability would likely require several tasks, including defining the expert panel program owner, standardizing expert panel staffing and deployment, and adopting a method of reporting that is agnostic to the panel topic.

CHAPTER FOUR

Recommendations from the Analysis of Alternatives

In the course of the AoA, we found that, although each option offered advantages, no single approach optimally addressed all aspects of an interoperability measurement system. As a result, we proposed a vision for a recommended Army interoperability measurement system, recommending an approach that combines elements of several alternatives analyzed. The recommended approach emerged from our discussions with the multinational fusion cell (MFC), our insights (gleaned both from other alternatives studied and decades of past experience), and comments from the interviewees. If implemented as intended, the new system will fulfill the Army's need for a standardized and robust methodology to measure interoperability levels achieved with partner nations during collective training events in a manner that enables it to develop solutions and communicate those solutions with its partners.

Conditions to Incorporate in the New System

Chapter Two outlined the dimensions we considered when analyzing the options. These dimensions include ease of use; cost incurred; consistency with, or similarity to, current Army processes; output relevance to stakeholders; balance of standardization and flexibility; reliability and sustainability; and differentiation from a readiness system. This section addresses each design dimension in light of the recommended new system.

Ease of use. Similar to CIRCuIT, the new system should be computer- or web-based. The system needs to be useable regardless of network connectivity. Completing the measures should be straightforward (such as T&EOs, 7ATC, or ABCANZ/MIAT) such that any observer with PFA subject-matter expertise can readily complete the measures. The measures should be as simple and one-dimensional as possible, similar to T&EO performance measures. This helps to reduce subjective judgments which, in turn, lessens the burden on observers and improves inter-rater reliability.

Cost incurred. To curb additional costs for measurement, the new system should leverage extant observer capabilities. This would include designing the system to be easily used by OC/Ts at Combat Training Centers and CAATs and ABCANZ lessons collection teams at exercises. The Army should strive to reduce any additional personnel resourcing solely for the purposes of measuring interoperability, which means that ease of use is critical to curtail costs. Data management and dashboard updates will require personnel, but as long as a computer-based collection tool includes automated data analysis capabilities, such as those available in R or other similar software applications, additional personnel should not be needed.

Consistency with, or similarity to, current Army processes. Measures in the new system should look very similar to those already collected during training events. We recommend that CIRCuIT matrices, provided by the CoEs, be deconstructed into a style more akin to ABCANZ CQL or Army T&EO performance measures, thereby creating the quantitative measures for the system. As a result, the measures would provide content and a style more similar to what the Army more widely uses, which would help ease the transition between systems. The system should also include a qualitative data component. This component is common in many options we analyzed and is needed for capability gap analyses. To have a qualitative component similar to what is already used, we recommend the Army adopt and, if necessary, make minor modifications to the method that CALL employs during its observations.

Output relevance to stakeholders. The Army wants a system that is robust, enduring, and capable of tracking changes over time—one that is based on a more quantitative performance measures list, such as T&EOs, could achieve this outcome; however, our AoA also highlighted the need for a qualitative element that can explain levels obtained and provide the Army with observations about why a specific level was achieved, how to improve it, and what might happen if no changes were made. We recommend that the Army have both a quantitative and a qualitative component, similar to CIRCuIT.[1] Specifically, we recommend that the qualitative component be designed based on CALL's format for inputting observations. The system must be firmly couched in Army doctrine to be fully accepted by those who collect data and use the results, including HQDA, Army Commands, Army Service Component Commands, CoEs, and Combat Training Centers. We recommend that the system be organized by PFAs; measures be nested in, but not organized by, ARTs; measures be developed and reviewed by CoEs; and the qualitative component be consistent with, and reviewed by, CALL. Furthermore, to make the output useful to stakeholders, the system should include an embedded analytic capability that automatically calculates interoperability levels by PFA, ties those levels to the qualitative data, and enables user-defined output presentation options.

Balance of standardization and flexibility. Our AoA identified a need to balance standardization and flexibility. Interviewees discussed a perceived need to have standardized data and results that can be analyzed across time and exercises. At the same time, the Army needs to be able to capture newly emerging challenges. To do all of this, the system needs to be flexible. We recommend that the new system have a quantitative component for capturing systemic trends, a component for capturing sufficiently detailed qualitative data to identify new issues and to conduct preliminary capability gap analysis, and a third component to perform automated data aggregation and analyses.

Reliability and sustainability. The new system needs to be reliable and sustainable across various exercises over varying time horizons. We recommend that the new system have a component with measures that are as straightforward as possible, directly map to interoperability, and are aligned to doctrine to foster universal understanding. To be sustainable, we recommend that the automated components of the system be based in a common software framework, such as Excel. Once the system is fully functional and time has passed with widespread adoption of the system, a more specialized software option could be considered.

---

[1] CIRCuIT's qualitative component is different than the approved CALL collection format.

Differentiation from a readiness system. The new system should make a clear distinction between measuring interoperability and assessing readiness. The Army already has a process in place for assessing readiness, as do most of its MN partners. The Army's process, which uses T&EOs for the training portion, is documented in AR 220-1, *Army Unit Status Reporting*.[2] The system we are recommending is not a replacement or an addition to this readiness assessment process, which is why it is important that users and stakeholders are educated on the differences between the proposed measurement system and a readiness assessment system. We are recommending a system that enables the Army to measure MNI levels obtained during training exercises to inform subsequent DOTMLPF-P decisions. This system should also be shareable with MN partners.

Next, we describe the proposed system.

Proposed System Overview

Purpose

The proposed system needs to provide interoperability levels by PFA to HQDA, Army stakeholders, and specified partners following signature training exercises. These levels must be tied to observations and recommendations across brigades, divisions, and corps. We recommend that the new system have three components: (1) an observer-based tool that facilitates the collection of systematic and standardized data, (2) a qualitative component that relies on additional input (e.g., from CALL) to explain variation in the observer-based tool output and allows flexibility to address senior leaders' interests that might not be addressed by a system's standardized elements, and (3) an embedded analytic tool expressed in the form of a user-selected dashboard.

[2] AR 220-1, *Army Unit Status Reporting and Force Registration—Consolidated Policies*, Washington, D.C.: Headquarters, Department of the Army, April 15, 2010.

Framework of the Three Components

The first component is an observer questionnaire with a yes, no, or select-one-response format.[3] As part of their typical prescribed role at signature training exercises, OC/Ts would observe training activities and interview training audiences. Based on information collected, they would complete Component #1 measures. We recommend that, when possible, observers review the system values that they record with the training audience.[4] We recommend that the quantitative measures that compose the observer questionnaire be nested in ARTs, be organized by PFAs, have a direct link to interoperability levels, and be developed and approved by the CoEs. We further recommend that the Army begin developing measures by deconstructing the 50 ART Level II task interoperability definitions already generated by the CoEs. These provide the language necessary for the measures, along with a direct objective connection to one of the four interoperability levels. In addition, all Component #1 measures should be a priori meta-tagged to Army WfF; the human, technical, and procedural aspects of interoperability; and ARTs. This meta-tagging will provide improved analytic filtering capabilities.

The second component would have virtually the same framework as the CALL approach, meaning that there would be explicit tags to DOTMLPF-P with detailed observation, discussion, recommendations, and implications provided. The CALL data collection format would need to have additional fields to enable connecting Component #1 and Component #2. These links will enable the user to conduct searches and analysis of the data. For instance, a user could see the level(s) for specific PFAs related to organization

[3] We recommend a fourth component: Collect metrics generated by tactical systems (e.g., number of minutes a network is available daily or during contact). This recommendation was not adopted by the Army, so we do not elaborate on it further in this report. The Army adopted a fourth component: an exploitation panel. This component is discussed in Chapter Five, where we describe the development of AIMS.

[4] Again, the proposed system is not a grade, assessment, or test. It measures the levels that partners are able to collectively reach during a training event. In addition, insights about why or how the training audience did what they did during training are very important to ensuring that the system's values are as accurate as possible.

and, thereby, see recommendations for improving interoperability in this domain.[5]

The third component would be a software-driven tool that automatically computes levels by PFA and enables user-defined automatic connection of Component #1 and Component #2 information. For early development, we recommend that the Army use software that is common across the Army and partner armies, such as Excel.

Strategic Support

HQDA DAMO-SSC's MFC is currently developing a set of interoperability road maps for a subset of its UAPs. These road maps specify macro-level training objectives for key activities between the UAP and Army forces. Road map components and outputs from the interoperability measurement system should be linked. System outputs should show the current level of interoperability, indicating whether units are on track to meet intermediate goals according to the road maps. Because of the system's ability to look at a specific measure within a PFA,[6] combined with the immense detail provided in Component #2, the system will help the MFC make adjustments to the road maps. The system and its employment should support senior decisionmakers and their staffs by triggering decision points and informing whether it may be necessary to reevaluate and reform interoperability goals and objectives.

Directing Measurement Requirements

We recommend that HQDA give direction to observers, specifying which aspects of interoperability (and, in turn, which Component #1 measures)

---

[5] Components #1 and #2 could be completed by the same group of observers or different observers. CALL observers have a long-standing operating principle of remaining nonattributional. Providing level values as in Component #1 may violate this principle. If so, we recommend that CALL observers complete Component #2 and other observers, such as OC/Ts, complete Component #1.

[6] We primarily recommended the Army significantly modify CIRCuIT because of likely observer confusion and thus measurement error; however, another advantage of the modification is that the AIMS user is able to see specific parts of the interoperability measure that need improvement.

will be observed and recorded. This can be done by specifying ART Level II tasks, PFAs, or specific subsets of measures within PFAs. We recommend that Component #1 measures be standardized and modified by HQDA and the CoEs only when changes in policy, doctrine, or organizations necessitate change. That being said, system collection requirements are likely to change by exercise scenario design, partner, and/or DOTMLPF-P changes, such as doctrinal changes or the fielding of new capabilities, equipment, or concepts.

Standardization and Flexibility

As just discussed, Component #1 measures should be standardized, with reviews happening at set intervals (possibly every 18 months, as is common with Army doctrine). Standardizing Component #2 is more difficult because it is primarily free-field entry; however, it has a standardized format for entry, with direction to observers to complete certain fields for every entry.

During our AoA, we saw a distinct need for the system to be flexible. For example, some training exercises may not fully stimulate all aspects of every PFA.[7] In such cases, those PFAs may not be measured, or only specific parts of the PFA would be measured. We recommend that the system achieve flexibility in three ways.

The most common means of achieving flexibility is with Component #2. Component #2 allows for free-field entry, and this is where unusual circumstances that affect interoperability levels should be recorded.

Less common should be the selection of Component #1 measures. Entire PFAs or measures within PFAs can be preselected prior to a training exercise. For major signature training exercises, we recommend that all PFAs and measures are employed, unless there is clear reason otherwise. For exercises, such as those done by an Army Service Combatant

[7] This is a common practice in training exercises. For example, senior trainers (the training audiences' senior commander) determine the primary training objectives prior to an exercise. There is often not enough time to train participants on everything, so some PFA elements may not be fully covered, and measurement of those elements would not be advised. Also, in some cases during an exercise, it might be determined that the training audience is strong in some areas and weak in others. Training emphasis then shifts to bolster weaker areas, leaving other areas not fully exercised.

Command or a Combat Training Center, in which training objectives parallel only a limited number of interoperability measures, then only training objective–relevant measures should be collected.

The final option for achieving system flexibility is revising Component #1 measures. However, we recommend that the first two options be pursued first and foremost. If these measures change with relative frequency (such as by exercise or quarterly), then users and decisionmakers will begin to lose confidence that the levels obtained over time represent valid trends. Also, if the measures change frequently, it will be more difficult to update, disseminate, and ensure version control of an automated standardized tool. If the revision process takes place, we recommend that multiple Army organizations—for example, multiple CoEs, HQDA, Army Futures Command, and other stakeholders—be part of the revision process. We recommend that the Army approach revisions to the system similarly to how it approaches making doctrine domain changes, as outlined by U.S. Army Training and Doctrine Command (TRADOC) Regulation 25-36, *The TRADOC Doctrine Publication Program*.[8]

Access to Data and Results

We did not develop a set of recommended policies and procedures for collecting, processing, and sharing data or disseminating results. We did, however, discuss with the client various considerations for where data and results should be stored, which we do not include here.

Soon after the AoA briefing, HQDA G-3/5 directed RAND Arroyo Center to have us shift our focus to support DAMO-SSC's development of the new measurement system. The Army named its new system AIMS. The Army decided that AIMS will have four components: (1) a quantitative observer-based instrument, (2) qualitative free-field entries, (3) an automated means to compute and display findings, and (4) an exploitation panel that synthesizes results to identify and take actions to resolve capability gaps. The next chapter provides an overview of the early development of AIMS.

[8] U.S. Army Training and Doctrine Command, TRADOC Regulation 25-36, *The TRADOC Doctrine Publication Program*, Fort Eustis, Va., May 21, 2014.

CHAPTER FIVE

Early Stages of Army Interoperability Measurement System Development

Because Component #1 (the measurement instruments) is new and a necessary precursor for Components #3 (the aggregation) and #4 (exploitation panel), early AIMS development focused on Component #1. Component #1 instruments were piloted at JWA 19 and TS 19.[1] During these exercises, Component #2–style data came from intact data collection processes previously developed for those exercises.[2] Component #3 was not developed or tested. Component #4 was piloted at JWA 19 but not at TS 19.[3] Following both JWA 19 and TS 19, HQDA produced records of decision (RODs)[4] between the United States and its UAPs summarizing interoperability levels and capability gaps associated with those levels.

The remainder of this report documents the development and testing of AIMS Component #1—the observer-based PFA measurement instruments. All PFA instruments were derived from CoE input that was provided to HQDA in support of CAA CIRCuIT development. Two versions of each PFA

---

[1] AIMS instruments were also distributed to some of the training audience, observers, and Army North representatives at Maple Resolve 19. Written and verbal feedback from them was collected and included in revisions to AIMS instruments.

[2] This report documents the development of AIMS. Because Component #2 was not created during the period of this report and no findings are included, we do not discuss this component.

[3] HQDA personnel were not in attendance at TS 19. Review and synthesis of AIMS findings occurred between U.S. and Australian representatives after the exercise.

[4] RODs are official memos that are signed by appropriate authorities to codify what was learned during an exercise, in this case, about interoperability.

instruments were piloted. The first pilot at JWA 19 yielded recommended changes from observers who completed AIMS instruments. Based on this pilot and feedback from Maple Resolve 19, we revised all instruments. These revised instruments were subsequently piloted during TS 19.

Component #1 Instrument Creation

We developed the initial version of the PFA measurement instruments based on CoE responses to an HQDA request. The request specified that the instruments provide interoperability level definitions for a specific set of ART Level II tasks.[5] Out of 50 total ART Level II tasks with interoperability levels defined (detailed in Appendix C), the Department of the Army specified 22 specific tasks for us to measure for the JWA 19 pilot. The remainder of this section uses "ART 3.2 Provide Fire Support" as an example, demonstrating the process of creating measures for the PFA instruments from individual ART Level II tasks.

For each ART, we first deconstructed the interoperability level definitions from the CoEs. Table 5.1 shows the original fires CoE definitions for ART 3.2. We deconstructed the ART definitions by identifying unitary constructs in each definition cell. In almost all cases, unitary constructs were defined as a sentence. In some cases, for compound sentences, we pulled out several unitary constructs from a single sentence. Table 5.2 shows the deconstruction of ART 3.2—all information is identical to Table 5.1, but the definitions for each interoperability level are broken down into constituent elements, which are more-manageable blocks.

Next, we took the deconstructed definitions and transformed them into questionnaire-style measures or interoperability measures. When possible, we created dichotomous go and no-go response formats for measures. For remaining measures where it was decided that a dichotomous format was not the most efficient, the response format asks the observer to select the option that best describes what they observed. For these multiple-choice options, we produced stem and root response formats, in which each root corresponded to a distinct interoperability level. Some stems already

[5] See Chapter Three for more about this.

had roots for all four interoperability levels (as shown in interoperability measures #1 through #3 in Table 5.3). For others, we had to create a 0 level.

TABLE 5.1

ART 3.2 Definitions of Levels of Interoperability

Level II ART	Integrated (I-3)	Compatible (I-2)	Deconflicted (I-1)	Not Interoperable (I-0)
ART 3.2: Provide Fire Support	U.S. Army and mission partners employ fires using organic digital capabilities that facilitate the collective and coordinated use of indirect, joint and electronic fires against surface targets, including nonlethal capabilities. The MPE network allows partners to send digital information in near real time that supports the D3A targeting process using each mission partner's organic fires C2 systems and support structure. U.S. Army and mission partners are both able to nominate and achieve electronic and computer network attack effects on targets. Little to no LNO support required.	U.S. Army and mission partners actively seek to establish interoperability by leveraging a combination of digital and analog capabilities or methodologies. Networked capabilities are limited to select digital-to-digital touch points often driving human intervention to conduct digital-to-analog-to-digital conversions with some LNO participation. The state of the MPE network requires mission partners to rely on some human dimensions to incorporate procedural controls in order to support the D3A targeting process. Liaison with U.S. network equipment required.	The mission partner operates with analog systems or leverages outdated, cyber-vulnerable digital capability. The mission partner relies heavily on liaison teams resourced with U.S. equipment and personnel often still requiring analog-to-digital conversions to support the D3A targeting process. The mission partner possesses little institutional knowledge or has limited MPE experience with coalition operations.	U.S. Army and mission partners do not Integrate fires using any digital, analog, or liaison means of communications. The mission partner possesses no institutional knowledge or MPE experience with coalition operations.

SOURCE: Task Definition Status provided by sponsor in March 2019.

TABLE 5.2
Deconstruction of ART 3.2

Level II ART	Integrated (I-3)	Compatible (I-2)	Deconflicted (I-1)	Not Interoperable (I-0)
ART 3.2: Provide fire support	U.S. Army and mission partners employ fires using organic digital capabilities that facilitate the collective and coordinated use of indirect, joint, and electronic fires against surface targets, including nonlethal capabilities.	U.S. Army and mission partners actively seek to establish interoperability by leveraging a combination of digital and analog capabilities or methodologies.	The mission partner operates with analog systems or leverages outdated, cyber-vulnerable digital capability.	U.S. Army and mission partners do not integrate fires using any digital, analog, or liaison means of communications.
D3A Targeting	The MPE network allows partners to send digital information in near real time that supports the D3A targeting process using each mission partner's organic fires C2 systems and support structure.	Networked capabilities are limited to select digital-to-digital touch points often driving human intervention to conduct digital-to-analog-to-digital conversions with some LNO participation.	They rely heavily on liaison teams resourced with U.S. equipment and personnel often still requiring analog-to-digital conversions to support the D3A targeting process.	The mission partner possesses no institutional knowledge or MPE experience with coalition operations.
	U.S. Army and mission partners are both able to nominate and achieve electronic and computer network attack effects on targets. Little to no LNO support required.	The state of the MPE network requires mission partners to rely on some human dimensions to incorporate procedural controls in order to support the D3A targeting process.	The mission partner possesses little institutional knowledge or has limited MPE experience with coalition operations.	
		Liaison with U.S. network equipment required.		

SOURCE: Task Definition Status, provided by sponsor March 2019.

TABLE 5.3
Stem and Root System of ART 3.2

ART 3.2 Interoperability Measures	I-Level
1. Select the one that best describes Army and MN partners' network in terms of employing fire support (select one).	
(a) Uses organic digital capabilities that facilitate the collective and coordinated use of indirect, joint, and electronic fires against surface targets to include nonlethal capabilities.	3
(b) Leverages a combination of digital and analog capabilities or methodologies. Networked capabilities are limited to select digital-to-digital touch points, often driving human intervention to conduct digital-to-analog-to-digital conversions with some LNO participation.	2
(c) Leverages analog systems or leverages with outdated, cyber-vulnerable digital capability.	1
(d) Does not integrate fires using any digital, analog, or liaison means of communications.	0
2. Select the one that best describes the conduct of D3A targeting process by U.S. and MN partners (select one).	
(a) Send digital information in near real time that supports D3A using each mission partner's organic fires C2 systems and support structure.	3
(b) Employ the MPE but rely on some human dimensions to incorporate procedural controls IOT support D3A.	2
(c) Require analog-to-digital conversions to support D3A.	1
(d) U.S. and MN partners did not integrate fires.	0
3. Select the one that best describes LNO involvement during fires integration (select one).	
(a) U.S. and MN partners successfully integrate fires with little to no LNO support necessary.	3
(b) Liaison with U.S. network equipment required.	2
(c) Rely heavily on liaison teams resourced with U.S. equipment and personnel.	1
(d) Did not establish liaisons: they operate independently.	0
4. U.S. and MN partners nominate and achieve electronic and computer network attack effects on targets.	3

SOURCE: RAND Arroyo Center modification of HQDA content.

For others still, the stem was missing a 1, 2, or 3 root.[6] Table 5.3 illustrates how the deconstructed definitions in Table 5.2 were transformed into a stem and root system with their corresponding interoperability levels.

Once interoperability measures were finalized for each ART Level II task, each set of ART task measures was placed into one of four PFA categories: CIS (which included IM/KM), intelligence, fires, or sustainment.[7] We created four PFA instruments to pilot at JWA 19.[8]

Observer-Based Tool Pilot 1: Joint Warfighting Assessment 19 and Maple Resolve

JWA 19 saw MN partner brigades integrated into a U.S. division, and AIMS observation teams comprised representatives from the United States, the UK, Canada, Australia, and New Zealand. Observers completed the instruments twice.

The first completion immediately followed a period during which units were conducting training in preparation for the main training event. This pre–main event completion provided us with valuable insights. We did not use the data from this first completion; instead, this first round was used as a trial period to familiarize and train observers on the instruments while also providing us with an opportunity to revise any glaring issues and refine the formulas used to aggregate interoperability measure values into a single PFA level result. Based on this first trial, we split the CIS instrument into two separate CIS and IM/KM instruments because different observers were observing these two general areas independently. For future

[6] We strove to stay true to the original wording from the CoEs, so we were conservative in making edits. In particular, other than creation of a 0-level root, we would not generate any missing 1, 2, or 3 roots.

[7] Assignment of ART task measure sets to PFAs were generally consistent with Army WfF. For example, all interoperability measures derived from the intelligence ART tasks went into the intelligence PFA instrument. However, measures derived from the movement and maneuver and protection WfFs were allocated to a PFA in a manner so that the observers of that PFA were likely to observe activities associated with that measure.

[8] See Appendix E for the computation of PFA interoperability levels.

implementation, we recommend that observers complete the instruments during the early stages of an exercise to develop familiarity with the measures and to make sure they are observing all aspects of the exercise needed to complete the instrument.[9]

The second completion occurred at the end of the main training event, prior to the final after-action reviews. The data from this collection were used to support the creation of RODs. Additionally, we met with each of the observation teams and collected their feedback on the instruments, and the feedback and comments were used to revise the instrument measures. These changes were tracked, and a record was maintained in audit tracking sheets, which were subsequently provided to the respective CoEs. Table 5.4 details the number of interoperability measures for each PFA, along with the number of recommended changes.

We piloted the JWA 19 versions of the intelligence and sustainment instruments at Maple Resolve 19. Here, U.S. sustainment and intelligence companies and teams were integrated into a Canadian brigade or Canadian service battalion, and there was a mix of U.S. and Canadian observers. Measures were completed once, at the end of the exercise. Overall, observers reviewed the measures and thought they were clear, well-written, and

TABLE 5.4

Number of Interoperability Measures Piloted at and Revised After Joint Warfighting Assessment 19

	Pilot 1		Pilot 2		
PFA	Measure Count	Recommended Changes Count	Measure Count	Recommended Changes Count	Not Battalion-Relevant
CIS	36	13	20	2	4
IM/KM			34	1	2
Intelligence	17	4	18	5	4
Fires	22	4	19	1	1
Sustainment	19	2	23	3	2

[9] Lt Col Kevin Taaffe from JMC recommended this trial completion to help train the observers. After the event, the RAND and JMC teams agreed that this approach likely improved data quality.

appropriate to focus on for interoperability. It was noted that there were other measures that could have been more effective to observe at the company level, such as certain tactical communications, shared ROEs, and compatibility of equipment and materiel (e.g., commonality in flat-rack haulers and fuel nozzles).

Takeaways from Joint Warfighting Assessment 19 and Maple Resolve 19

Piloting instruments at JWA 19 was crucial in the development of AIMS. First and foremost, the pilot demonstrated how AIMS can combine quantitative and qualitative data to support an exploitation panel's synthesis of lessons into capability gaps that will inform the United States and UAP about needed improvements to yield higher levels of interoperability. These were documented in the exercise RODs.

The pilot also provided an important opportunity to revise AIMS instruments and improve future AIMS implementation. Splitting the CIS PFA into one CIS instrument and one IM/KM instrument improved the ease of completion of those measures. In the subsequent version of the AIMS tool, CIS and IM/KM are separated. Of the 36 measures that were in the original CIS PFA measurement instrument for JWA 19 and Maple Resolve 19, 12 remained in CIS, while 24 migrated into the IM/KM PFA instrument.

We also received feedback from observers about individual interoperability measures. Across the PFAs, observers recommended changes to 24 percent of all interoperability measures. As noted in Table 5.4, CIS had the most recommended changes, with 13 of the 36 measures being noted as needing change. Observers for fires and intelligence recommended changes to four measures. Sustainment observers recommended changes to that only two of the 19 total measures. The nature of the recommended changes varied across the PFAs as well. Frequent notes recorded were "bad wording" or "too broad" for specific measures, and these often referred to measures with unclear phrasing and/or measures with unclear echelons. Some observers had stronger recommendations, including rewriting a measure or removing it in its entirety from the instrument. In other instances, observers recommended including doctrine as a footnote or citation for the measure.[10]

We also received feedback beyond the instruments. Several observers noted a need for a standardized, rigorous, and clear set of instructions. The set of instructions should include general background on the way that interoperability was being measured, how the measures were derived, and doctrinal references for measures that may not be well known.

Feedback in the form of comments and clarifications about the measures from JWA 19 and Maple Resolve 19 was used to revise measures that would be piloted at TS 19.[11]

Observer-Based Tool Pilot 2: Talisman Sabre 19

Prior to employing the post–Pilot 1 revised PFA instruments at TS 19, we went back to the ART Level II tasks that had not been considered for inclusion in the original PFA instruments. We did this with the intention of identifying other measures that would be beneficial to add to the instruments. We included measures from 15 additional ART Level II tasks during this step.[12] These are indicated in Appendix C.

At TS 19, there was a RAND researcher on the ground working with the observers.[13] Like JWA 19, observers were part of an MN observation team and were predominantly from Australia, with some New Zealand and U.S. personnel present as well. All five PFA instruments were tested bilaterally at two levels—a U.S. brigade combat team integrated into an Australian

[10] All these recommendations were cataloged and shared with the respective CoEs to aid in their refinement of their AIMS instruments.

[11] We did not incorporate all changes into the revised instruments because we were conservative in the extent of wording changes performed to the original CoE input. However, the feedback and changes we recommended were provided to the HQDA and the CoEs.

[12] Because of concerns about instrument length, we did not include all measures that would have been indicated during our deconstruction of the CoEs' original definition input.

[13] This included distributing instruments, interviewing observers and training unit personnel, and reviewing observation logs.

Combined Joint Task Force, and a U.S. infantry battalion integrated into an Australian brigade. For the latter, observers were specifically from the Joint Pacific Multinational Readiness Capability. All observers received the instruments at the end of the training exercise and had not seen them prior. They were directed to do two things: (1) complete the instrument for the exercise and (2) provide feedback on individual measures. For the battalion-level observers, there was a third task—observers were to indicate whether they thought an individual measure of performance was germane to a battalion integrated into a brigade, even if that measure of performance was not observed during the exercise. Table 5.4 details the number of interoperability measures for each PFA, along with the number of recommended changes from this exercise and the number of measures that were considered irrelevant to battalions integrated into a brigade (Pilot 2).

Takeaways from Talisman Sabre 19

Compared with JWA 19, there were fewer suggested measurement revisions from TS 19 (see Table 5.4), and the suggested changes received were generally less extensive. The intelligence PFA requires the greatest percentage of changes (five out of 18, or 27 percent). The total number of recommended changes to CIS and IM/KM following TS 19 (three) was much lower than the changes recommended during JWA 19 (13).

We also solicited observers' opinions about ease of completion and overall value of the measurement. The observers, notably the Australian members, were keenly supportive of the measurement process and indicated a desire to incorporate parts of the AIMS instruments into processes after the exercise. It was noted that observers rapidly completed about 90 percent of the measures, while the other 10 percent would have been easier to complete if observers had prior knowledge of the measures. That is, during the exercise, the observers had not looked at the activities related to a small proportion of the measures. Had the observers known in advance, this final 10 percent of measures also would have been easy to complete rapidly. This reinforces the lessons from JWA 19: It is valuable for observers to be familiar with the measures at the start of the exercise.

Joint Pacific Multinational Readiness Capability OC/Ts completed and provided feedback on the PFA instruments with reference to the U.S. battalion being integrated into the Australian brigade. AIMS instruments were intended for a brigade corps, so we asked OC/Ts to comment on completing them for battalions. They felt confident about being able to complete most measures on the instruments (see Table 5.4), although they felt that several would require some revision to be better suited for battalion-level measurement.

CHAPTER SIX

Concluding Remarks

The Army is undertaking significant changes to how it prepares to interoperate with its critical partners. These changes include revising doctrine, creating a new Interoperability Campaign Plan, and developing strategic road maps with specific partners to monitor and track changes to achieved interoperability levels. To support the management and tracking of interoperability gains, the Army identified a need for a system for measuring interoperability and pursued the development of such a system.

The Army wants an enduring system that leaders and staff can use to look at trends over time and to organize training events. They want the system to inform the entire interoperability enterprise (that of both the United States and partners) about capability gaps that are limiting achievement of projected levels of interoperability.

In this report, we summarize the identification of characteristics (derived from the dimensions just mentioned) associated with a useful and relevant measurement system and present the preliminary efforts conducted to develop the system.

Key Characteristics That the System Should Have

Ease of use. The new system should be computer- or web-based. It needs to be useable, regardless of network connectivity. Completing the measures should be straightforward; this helps reduce the amount of subjective judgment required, which, in turn, lessens the burden placed on observers and improves inter-rater reliability.

Cost incurred. The Army should strive to reduce any additional personnel resourcing for the sole purpose of measuring interoperability.

Therefore, ease of use is important to curtail costs. To curb additional costs for measurement, the new system should leverage extant observer capabilities, including OC/Ts, CAATs, and ABCANZ lessons collection teams already present at exercises.

Consistency with, or similarity to, current Army processes. Measures in the new system should look very similar to those that are already collected during training events. The system also should include a qualitative data component needed for capability gap analyses.

Output relevant to stakeholders. We recommend that the system have both a quantitative and a qualitative data component with an embedded analytic capability that automatically calculates interoperability levels by PFA, ties levels to the qualitative data, and provides user-defined output to enable capability gap analysis.

Balance of standardization and flexibility. The system should have (1) a standardized format for quantitative data that allows them to be analyzed over time and across exercises and (2) a flexible format for qualitative data to capture newly emerging challenges. The system should also make available an ability to combine the two and analyze any discrepancies.

Reliability and sustainability. The system should have a component with measures that are as straightforward as possible, directly map to interoperability, and are aligned with doctrine to foster universal understanding. To be sustainable, the automated components of the system should be based in a common software framework, such as Excel. Once the system is fully functional and time has passed with widespread adoption of the system, a more specialized software option could be considered.

Differentiation from a readiness system. The Army should develop a measurement, not an assessment system, and work to make sure that users and stakeholders are educated on the differences.

Army Interoperability Measurement System

After our AoA, the Army decided to develop a new system for measuring MN interoperability—the AIMS. The Army further decided that AIMS would include many of the characteristics identified in the AoA—namely, a quantitative instrument for measuring interoperability levels, a qualitative component to enable capability gap analysis, and an automated approach to connect and analyze the data. The Army also incorporated a fourth component, exploitation panels. These panels would convene immediately following a training exercise and comprise representatives from all participating countries. Exploitation panels collectively work to ensure that the results derived from AIMS Component #1 are consistent with ground truth, and, if there is an issue, they adjudicate it. Exploitation panels play a critical role in synthesizing the results from the quantitative and qualitative data analysis to identify capability gaps and take action(s) to resolve them.

Further refinement and development of AIMS is ongoing; however, it appears that its development is progressing well. AIMS was used to formulate several records of decisions following JWA 19 and TS 19 that highlighted capability gaps. AIMS instruments are both similar and dissimilar to prior Army measurement approaches; however, to date, observers who have completed the instruments rarely, if ever, had issues completing the instruments.

APPENDIX A

Analysis of Alternatives Questions List

Prior to conducting interviews with system proponents from each of the options analyzed, we developed a list of questions that were considered important to determining the origin of the assessment, the logistics of implementation and data analysis, correlation to interoperability levels, ability to support senior decisionmakers and partners, reliability, and sustainability, among other things. A full list of the questions is in this appendix. Prior to interviews, we sent this list to the interviewees, as described in the body of this report, to give them an opportunity to develop answers in advance and to guide the interviews.

Assessment measures:

- What were the foundational concepts for developing this methodology?
- Describe the various elements/aspects of this methodology.
- How many items are there to complete?
- How is the rating scale set up? What is the rating scale? (binary, categorical, text, etc.)

Assessment implementation:

- Who implements the assessment?
- How is the assessment implemented?
- Is self-assessment permitted?
- What aspects of the assessment are or are not required?

- How often do assessments occur? Are they regular, synchronized with deployment schedules, only conducted when directed?
- How much does implementation cost?
- How easy is implementation?

Collection methods:

- Who collects the data/information?
- How are the data/information collected?
- How standardized are collection measures?
- At which echelon are data collected?
- What is the cost of collection?
- How easy is collection, data/information capture?
- What is the collection modality? (Subjective/objective, quantitative/qualitative)

Type of data captured:

- Are dimensions of readiness measured/captured? If so, explain.
- Are the data captured numeric, verbal, or some combination?

Type of output:

- Is the output quantitative, qualitative, or some combination?
- What framework is the output produced within? (For example, output within AR 34-1 framework.)
- What is the degree of standardized output? (Are results from this assessment directly comparable with outputs from other assessments?)

Data processing:

- To what degree is data processing automated?
- Who controls the data processing?
- What types of data linkages (ABCANZ LOE [line of effort], WfF, PFAs, etc.) currently exist?

Analytics:

- To what degree is the analytics step automated?
- Who controls the analytics step?
- How flexible is the system to make changes?

Ownership of data:

- Who owns the data?

Access to data or data output:

- Who has access to the data?
- At what level of access does the data reside? (Classification)
- Who has access to the data output?
- At what level of access does the output reside? (Classification)

Support of strategic assessments of interoperability:

- Which parts of the road map are supported?
- How close is the relationship between a road map component and assessment outputs?
- Are measures directly/indirectly connected to DOTMLPF-P?
- Are MOPs [measures of performance]/MOEs [Measures of effectiveness] nested within the assessment? If not, is there a direct/indirect way to map the assessment to MOPs/MOEs?
- Does output directly correlate to interoperability level? Or is an intermediate step needed to get from assessment output to interoperability level?

Support of other entities and partners:

- How readily does it, or could it, support senior decisionmakers?
- How readily does it support our partners?
- Will it effectively support/inform Army CoE and CFT [cross-functional team] activities?

Ease of adoption by the training enterprise and the field:

- How readily can this be adopted by the training enterprise and the field?
- How readily is this assessment integrated with DTMS [digital training management system]?
- How readily is this assessment integrated with CUSR [commander unit status report]?
- How easy is it to employ during exercises?

Face validity of measures:

- How capable is the assessment of capturing likely limitations and possible solutions?
- Are there known and/or reported workarounds or cheats?
- What is the existing buy-in from units? From those who use these data?
- Is the assessment sustainable? Statistically robust?

Likely reliability of measures and total scores:

- Is there good inter-rater reliability?
- Is there a regular review of the assessment to ensure currency and validity?

Standardized and repeatable across time:

- Is the system amenable to being used across multiple exercises and years with minimal modification?
- Does it produce results that can be compared across the different samples?

APPENDIX B

Completed Questions List for All Considered Alternatives

There were typically two rounds of interviews conducted with proponents from each of the alternatives.[1] Prior to the first interview, we sent a partially answered list of all the questions. These answers were derived from review of material obtained prior to the interview. After the initial interview, the spreadsheet was filled in based on our synthesis of interviewee responses as well as our prior document review. Questions that were lacking answers were highlighted, and this updated list of questions and answers was sent back to the interviewees to (1) confirm that the existing answers filled in were correct and (2) prepare them for a second interview in which we hoped to answer the remaining questions. After the second and final interview, the finalized question list was sent to the proponents from each approach to ensure accuracy before including in this report. This appendix presents the complete lists of questions and approved responses for each alternative analyzed. Responses to questions are italicized. Questions that were not answered are not included. Questions denoted with an asterisk indicate responses that we filled in that were not confirmed by the proponents. Note that we removed answers to questions about data location and ownership, per guidelines.

[1] For some options, the second interview was deemed unnecessary because the answers to the research questions were complete and accurate. For other representatives, more than two interviews occurred to provide a better understanding of the options.

Center for Army Lessons Learned

Assessment measures:

- What were the foundational concepts for developing this methodology? *Responsive to the operational force; responsive to the institutional force; covers all levels of war; covers all warfighting functions; includes unified action partners; supports a fully integrated lessons sharing culture; is sustainable.*
- How many items are there to complete? *Undefined. Item completion and write-ups will be determined by the number of unique observations or lessons observed that seem notable to the observer team.*
- How is the rating scale set up? What is the rating scale? (binary, categorical, text, etc.) *Text, written prose.*

Assessment implementation:

- Who implements the assessment? *A team of observers sent from the Center for Army Lessons Learned.*
- How is the assessment implemented? *Observations made over the course of an exercise are combined in a final write-up, in written prose, and provided to the observed unit. Changes and adherence to the lesson learned is at the discretion of the observed unit's commander.*
- What aspects of the assessment are or are not required? *Written observations are required, but if nothing unusual is noted, positive or negative, lessons learned are not compulsory.*
- How often do assessments occur? Are they regular, synchronized with deployment schedules, only conducted when directed? *Observations written into a CALL product are only as regular as the Army's exercise tempo is. If exercises are less frequent, then CALL products will also be less frequent. Since CALL has to observe a unit in action (live virtual or constructed), a unit must be in one of these circumstances to be able to generate an assessment/observation.*
- How much does implementation cost? *CALL is funded independent of the evaluated unit, but individual exercise budgets would fund the CALL team's ability to be present at exercises. A general estimate for a*

single CALL personnel to attend a one- to two-week exercise observation averages US $4,000, mainly for travel expenses.

- How easy is implementation? *CALL believes implementation is relatively simple. Their personnel are trained for observation, embed with units, and passively observe with some or limited interaction with teams to generate context for their lessons learned products.*

Collection methods:

- Who collects the data/information? *CALL observer teams—could be 10–40 people, depending on the observation event.*
- How are the data/information collected? *Observers join an exercise; information is collected from direct visual observation of training environment and recorded on hardcopy documents that are later converted and edited into softcopy text.*
- At which echelon are data collected? *All echelons.*
- What is the cost of collection? *(See above.)*
- How easy is collection, data/information capture? *Relatively easy.*
- What is the collection modality? (Subjective/objective, quantitative/ qualitative) *Subjective, qualitative.*

Type of data captured:

- Are dimensions of readiness measured/captured? If so, explain. *Readiness observations are generated, but these are not conveyed as metrics, only as notable observations, perhaps with recommendations as to the cause or corrective course of action.*
- Are the data captured numeric, verbal, or some combination? *Verbal and visual data captured in written prose.*

Type of output:

- Is the output quantitative, qualitative, or some combination? *Qualitative; text and written prose.*
- What framework is the output produced within? (For example, output within AR 34-1 framework) *Output guided according to AR 11-33.*

- What is the degree of standardized output? (Are results from this assessment directly comparable with outputs from other assessments?) *Output for each event is consolidated written prose. There is a high degree of similarity between CALL products, but emphasis and lead writer can change from product to product.*

Data processing:

- To what degree is data processing automated? *It is not automated or analytic.*
- Who controls the data processing? *N/A*

Analytics:

- To what degree is the analytics step automated? *It is not automated or analytic.*
- Who controls the analytics step? *N/A*
- How flexible is the system to make changes? *Flexible for observation changes and emphasis according to unit, CoEs, etc.*

Support of strategic assessments of interoperability:

- Which parts of the road map are supported? *Unknown.*
- How close is the relationship between a road map component and assessment outputs? *Unknown.*
- Are measures directly/indirectly connected to DOTMLPF-P? *According to AR 11-33 and CALL interviews, measures are directly connected to DOTMLPF-P.*
- Are MOPs/MOEs nested within the assessment? If not, is there a direct/indirect way to map the assessment to MOPs/MOEs? *They can be, particularly if the unit or CoE is looking to observe something specific about a MOP for a particular exercise. MOP assessment would not be considered a compulsory item in CALL observations.*
- Does output directly correlate to interoperability level? Or is an intermediate step needed to get from assessment output to interoperability level? *Output is in written prose and does immediately*

correlate to interoperability level. A subjective crosswalk would be required to achieve this end.

Support of other entities and partners:

- How readily does it, or could it, support senior decisionmakers? *It can readily be used to support senior decision leaders (it currently supports senior decisionmakers).*
- How readily does it support our partners? *CALL is actively used in UAP and NATO settings. It can readily be used to support partners.*
- Will it effectively support/inform Army CoE and CFT activities? *It will effectively support/inform Army CoE and CFT activities. CoEs and CFTs are sometimes joined with CALL in observation efforts.*

Ease of adoption by the training enterprise and the field:

- How readily is this assessment integrated with DTMS? *Unknown.*
- How readily is this assessment integrated with CUSR? *Unknown.*
- How easy is it to employ during exercises? *Relatively easy.*

Face validity of measures:

- How capable is the assessment of capturing likely limitations and possible solutions? *Very capable. CALL provides context, anecdotal evidence specific to a unit, and often reflects direct interaction with the observed unit to identify solutions and limitations.*
- Are there known and/or reported workarounds or cheats? *A unit has full discretion to comply or respond with CALL observations. Since CALL links to warfighting functions, its likely units will choose to respond or ingest training guidance. On face though, CALL observations are not compulsory for units.*
- What is the existing buy-in from units? From those who use the data? *Anecdotally, buy-in is high, especially with units about to deploy for overseas combat operations—CALL products are well known in Army (and joint) culture.*
- Is the assessment sustainable? Statistically robust? *Assessments are sustainable in the sense that observers have been historically present*

for decades and will continue to be present at major Army training evolutions.

Likely reliability of measures and total scores:

- Is there good inter-rater reliability? *Probably not.*

Standardized and repeatable across time:

- Is the system amenable to being used across multiple exercises and years with minimal modification? *Observations are repeatable in the sense that CALL is already set to an exercise tempo for almost every major Army evolution. CALL is not standardized in the same sense as metric recording is—different lead authors for the report will have different writing styles (although still similar), and different observation teams will report and observe different things.*
- Does it produce results that can be compared across the different samples? *No.*

RAND Arroyo Center Approach Developed for the 7th Army Training Command

Assessment measures:

- What were the foundational concepts for developing this methodology? *Based in Army's AUTLs* [Army universal task list], *METs, and SCTs for a BCT and input from lessons learned, taking elements from the T&EO system.*
- Describe the various elements/aspects of this methodology. *Many of the items are T&EO items exactly as written in AUTL—other times, they are items from lessons learned that were crafted into a language similar to T&EOs. So the questions are all qualitative but with a numeric response format, and when T&EOs are scored, the scores are automatically linked to dimensions.*

- How many items are there to complete? *80.*

- How is the rating scale set up? What is the rating scale? (binary, categorical, text, etc.) *Go/no-go*

Assessment implementation:

- Who implements the assessment? *7ATC*
- How is the assessment implemented? *Data are collected by OC/Ts, then given back to RAND, where data are put into an automated dashboard.*
- Is self-assessment permitted? *No.*
- What aspects of the assessment are or are not required? *You complete the measures that are relevant to your team, whether it's maneuver, fires, intel, brigade staff training, etc.*
- How often do assessments occur? Are they regular, synchronized with deployment schedules, only conducted when directed? *You can assess any time after a significant portion of an exercise is complete, any time after you've had enough observation time.*
- How much does implementation cost? *It's very little additive cost because you already have intact OC/Ts on the payroll, and they'd already be looking at the elements this assessment is asking about, so they're not being asked to do something super new. The only new aspect is completing the measures.*
- How easy is implementation? *Very.*

Collection methods:

- Who collects the data/information? *Data are collected by OC/Ts. Currently, OC/Ts would provide RAND with data, and we would put it into an automated R programming dashboard. Long term, the idea is that 7ATC would be in control of every step of the process.*
- How are the data/information collected? *It's all observation based.*
- How standardized are collection measures? *Very.*
- At which echelon are data collected? *System is built specifically for brigade, but you could apply it to battalion as well.*
- What is the cost of collection? *(See above.)*
- How easy is collection, data/information capture? *Very.*

- What is the collection modality? (Subjective/objective, quantitative/qualitative) *The questions are all qualitative, but with a numeric response format.*

Type of data captured:

- Are dimensions of readiness measured/captured? If so, explain. *It's not a measurement of readiness. Some call it a measure of "interoperability readiness," but that's not an approved readiness variable within the Army system.*
- Are the data captured numeric, verbal, or some combination? *Combination.*

Type of output:

- Is the output quantitative, qualitative, or some combination? *The assessment provides a T 1/2/3/4 level of response, which is like Objective T.*
- What framework is the output produced within? (For example, output within AR 34-1 framework) *Objective T.*
- What is the degree of standardized output? (Are results from this assessment directly comparable with outputs from other assessments?) *Once measures are determined, they'd be very standardized and fairly invariant, only changing if the Army changes its doctrine.*

Data processing:

- To what degree is data processing automated? *Once the data are input into the R dashboard, then there's more automation. But the assessment is still very human-dependent.*
- Who controls the data processing? *7ATC.*
- What types of data linkages (ABCANZ LOE, WfF, PFAs, etc.) currently exist? *WfF; human, procedural, and technical dimensions; HQDA G-3/5/7 IO dimensions; easy to add additional linkages.*

Analytics:

- To what degree is the analytics step automated? *(See above.)*
- Who controls the analytics step? *7ATC.*
- How flexible is the system to make changes? *Very.*

Support of strategic assessments of interoperability:

- Which parts of the road map are supported? *Not designed to support road map, as the road map did not exist when the system was built. There's potential for it to support road maps, but unsure.*
- How close is the relationship between a road map component and assessment outputs? *N/A*
- Are measures directly/indirectly connected to DOTMLPF-P? *No.*
- Are MOPs/MOEs nested within the assessment? If not, is there a direct/indirect way to map the assessment to MOPs/MOEs? *They're not. In a sense, if your MOE was WfF, or human, you could roll up. But this was not built into the system.*
- Does output directly correlate to interoperability level? Or is an intermediate step needed to get from assessment output to interoperability level? *Yes—no intermediate step needed.*

Support of other entities and partners:

- How readily does it, or could it, support senior decisionmakers? *It would require an additional layer of analysis to support senior decisionmakers. The system can provide problem areas, in terms of WfF or human/procedural/technical, but it does not readily prescribe/identify any capability gap or potential solution. This is perhaps the biggest shortcoming of the system.*
- How readily does it support our partners? *It wasn't designed to—[it was] designed to enable 7ATC to communicate with partners. So, 7ATC leadership and JMRC leadership would be able to show partners trends, but that is about it.*
- Will it effectively support/inform Army CoE and CFT activities? *Similar to DOTMLPF, this would require another level of analysis. The system simply provides you with the problem area. Because this system*

directly links to WfF, that does streamline the ability to communicate with CoEs since CoEs are split out by WfF.

Ease of adoption by the training enterprise and the field:

- How readily can this be adopted by the training enterprise and the field? *Easily, as it is consistent with how U.S. forces measure their own training already.*
- How readily is this assessment integrated with DTMS? *N/A*
- How readily is this assessment integrated with CUSR? *N/A*
- How easy is it to employ during exercises? *Very.*

Face validity of measures:

- How capable is the assessment of capturing likely limitations and possible solutions? *It captures limitations/problem areas, but not solutions.*
- Are there known and/or reported workarounds or cheats? *No.*
- What is the existing buy-in from units? From those who use the data? *Very high.*
- Is the assessment sustainable? Statistically robust? *Sustainable—yes. Statistically robust—unsure.*

Likely reliability of measures and total scores:

- Is there good inter-rater reliability? *Unsure.*
- Is there a regular review of the assessment to ensure currency and validity? *No, but (1) the measures would really only change if Army doctrine changed, and (2) it is an easy system to update/change, so staying current and valid is not a tremendous concern.*

Standardized and repeatable across time:

- Is the system amenable to being used across multiple exercises and years with minimal modification? *Yes.*
- Does it produce results that can be compared across the different samples? *Yes.*

Approach Employed by Joint Modernization Command at Joint Warfighting Assessment

Assessment measures:

- What were the foundational concepts for developing this methodology? *Priority focus areas from HQDA EXORD 293-17, MNI Training.*
- Describe the various elements/aspects of this methodology. *Beginning with PFAs, we derived questions to address human, procedural, and technical dimensions. Analysis identifies actionable gaps and associated DOTMLPF-P solutions.*
- How many items are there to complete? *Five overarching questions, 40 subquestions.*
- How is the rating scale set up? What is the rating scale? (binary, categorical, text, etc.) *Categorical, text.*

Assessment implementation:

- Who implements the assessment? *We implement it with a team of about 40 individuals who perform the assessment. There are also internal assessors who get after the technical component. Because JMC has limited resources, we're reliant on externals, so the Army and ABCANZ both bring in their own members.*
- How is the assessment implemented? *Observation based. This assessment is not focused particularly on interoperability, but they'll get some interoperability discussions from everything else.*
- Is self-assessment permitted? *At a high level—yes.*
- What aspects of the assessment are or are not required? *We assume all developed measures will be addressed and if observations can be made and recorded, they will be.*
- How often do assessments occur? Are they regular, synchronized with deployment schedules, only conducted when directed? *We assume daily hotwashes among observer teams. But we are unsure how many data collections will occur during JWA.*
- How much does implementation cost? *Need an estimate of the number of personnel specifically devoted to the interoperability assessments.*

- How easy is implementation? *Not easy at all—it's both labor- and time-intensive.*

Collection methods:

- Who collects the data/information? *The team.*
- How are the data/information collected? *Observations, focus groups, and surveys.*
- How standardized are collection measures? *They're tailored to the event, so there lacks a level of cohesion from one event to the next.*
- At which echelon are data collected? *CJTF [Combined Joint Task Force] and below, down to brigade level.*
- What is the cost of collection? [No answer.]
- How easy is collection, data/information capture? *Similar to an OC/T-like construct. Relies on expert observers who have good access to view training activities.*
- What is the collection modality? (Subjective/objective, quantitative/qualitative) *All of the above, depending on what information you're collecting. Fire mission accuracy is both qualitative and objective. Focus groups are qualitative and subjective. Fire mission time lag is quantitative and objective. Text analytics are quantitative and subjective.*

Type of data captured:

- Are dimensions of readiness measured/captured? If so, explain. *Focused on interoperability, not readiness.*
- Are the data captured numeric, verbal, or some combination? *Combination.*

Type of output:

- Is the output quantitative, qualitative, or some combination? *Combination.*
- What framework is the output produced within? (For example, output within AR 34-1 framework) *Each assessment leverages best practice of standards-based assessments using same standard of AR 34-1.*

- What is the degree of standardized output? (Are results from this assessment directly comparable with outputs from other assessments?) *Existing assessments are not fully comparable. Although all agencies use the same rating scale, they rate different categories, from the warfighting functions to the warfighting challenges. Exercises also differ by participating country, scenario, and size of units (BN [battalion], BDE [brigade], DIV [division]). It is not a problem solved by governance, as each agency will continue to assess for their own specific purposes.*

Data processing:

- To what degree is data processing automated? *Conducted on a computer, but very human dependent.*
- What types of data linkages (ABCANZ LOE, WfF, PFAs, etc.) currently exist? *PFAs.*

Analytics:

- To what degree is the analytics step automated? *None of it is automated.*
- Who controls the analytics step? *JMC owns the analytics step.*

Support of strategic assessments of interoperability:

- Are measures directly/indirectly connected to DOTMLPF-P? *Yes, recommendations are tied to DOTMLPF-P.*
- Does output directly correlate to interoperability level? Or is an intermediate step needed to get from assessment output to interoperability level? *Yes—directly correlated to interoperability level.*

Support of other entities and partners:

- How readily does it support our partners? *Very readily—implement ABCANZ/NATO/other agreed-upon multinational standards in assessment. Objectives include enabling MN partners to evaluate concepts and capabilities as well.*
- Will it effectively support/inform Army CoE and CFT activities? *Yes.*

Face validity of measures:

- Is the assessment sustainable? Statistically robust? *No.*

Standardized and repeatable across time:

- Is the system amenable to being used across multiple exercises and years with minimal modification? *Yes.*
- Does it produce results that can be compared across the different samples? *Yes.*

CIRCuIT Framework

Assessment measures:

- What were the foundational concepts for developing this methodology? *Combined ART level tasks with AR 34-1 interoperability levels.*
- Describe the various elements/aspects of this methodology. *Observer (planned to be self-assessed) selects the desired ARTs. Based on observation of training behavior, the observer records the interoperability level (0–3). The levels are loaded into a web-based app that also has the user select several variables that describe the training event. In addition, the user is able to select DOTMLPF-P variables that are related to the solution/limitation related to the observed interoperability level.*
- How many items are there to complete? *There are 50 ARTs measures, but observers only fill out a small selection of these based on their interest.*
- How is the rating scale set up? What is the rating scale? (binary, categorical, text, etc.) *Categorical, with four score options.*

Assessment implementation:

- Who implements the assessment? *Self, or other.*
- How is the assessment implemented? *Web-based.*
- Is self-assessment permitted? *Most data would be self-assessed; definitions of levels are the key mechanism to reduce bias.*

- What aspects of the assessment are or are not required? *User selected but could be directed.*
- How often do assessments occur? Are they regular, synchronized with deployment schedules, only conducted when directed? *Any Army event with the United States and joint/MN partners. Deployments and collective training.*
- How much does implementation cost? *Cost would be relatively low because observers are either unit personnel or observers of the training event.*
- How easy is implementation? *Not difficult for a trained observer.*

Collection methods:

- Who collects the data/information? *CALL/user collects/cloud solution storage that is provided by HQDA G-8.*
- How are the data/information collected? *Unit self-collected or through trained exercise observers.*
- How standardized are collection measures? *The 50 ART anchors will be standardized, but the values obtained may not be because the observer must make a subjective judgment about the levels in cases when all of the subelements of the scale values are not observed or a combination of subelements across levels are observed.*
- At which echelon are data collected? *Not echelon dependent; it does appear that the scale values subelements are largely for echelons above brigade.*
- How easy is collection, data/information capture? *Fairly straightforward. Once trained and with the web app, it should be relatively easy.*
- What is the collection modality? (Subjective/objective, quantitative/ qualitative) *Combines subjective and objective. Has an objective tool. Users can provide qualitative input related to the levels recorded with the tool.*

Type of data captured:

- Are dimensions of readiness measured/captured? If so, explain. *The values collected with the tool could be developed into a readiness value.*

- Are the data captured numeric, verbal, or some combination? *Combination.*

Type of output:

- Is the output quantitative, qualitative, or some combination? *Combination.*
- What framework is the output produced within? (For example, output within AR 34-1 framework) *AR 34-1.*
- What is the degree of standardized output? (Are results from this assessment directly comparable with outputs from other assessments?) *Possibly, but not clear.*

Data processing:

- To what degree is data processing automated? *Fairly automated; additional automation could be developed.*
- Who controls the data processing? *CAC* [Combined Arms Center] *interoperability proponents or maybe SSC* [Soldier Systems Center], *not fully sure.*
- What types of data linkages (ABCANZ LOE, WfF, PFAs, etc.) currently exist? *Relational data base, echelon, observer, mission, which COCOM [combatant command] is supported, type of operations, solutions/limitations, which network is being used, DOTMLPF-P, bilateral/multilateral is computed and then included in the database.*

Analytics:

- To what degree is the analytics step automated? *User has assessment view tabs, with reactive pull downs, mission partner, echelons by the United States and MN partner, could do by PFA category, bar plots for solutions by DOTMLPF-P. Long-term goal is to have text analytics built-in. Can display the raw data if requested.*
- Who controls the analytics step? *CAA at this point, but it would be the MFC.*
- How flexible is the system to make changes? *Simple, written in R.*

Support of strategic assessments of interoperability:

- Which parts of the road map are supported? *Most likely the objectives.*
- How close is the relationship between a road map component and assessment outputs? *Amenable to close connection. But it is dependent on several factors: if MFC directs the ARTs selected and DOTMLPF-P data collected.*
- Are measures directly/indirectly connected to DOTMLPF-P? *Complete the ART, then drop down IE level, then detail about the IE.*
- Are MOPs/MOEs nested within the assessment? If not, is there a direct/indirect way to map the assessment to MOPs/MOEs? *MOPs (Interoperability level subelements) are woven into the MOE (interoperability levels).*
- Does output directly correlate to interoperability level? Or is an intermediate step needed to get from assessment output to interoperability level? *Directly correlates to levels.*

Support of other entities and partners:

- How readily does it, or could it, support senior decisionmakers? *It could show where units stand on interoperability. It would depend on the decisionmakers' questions.*
- How readily does it support our partners? *Knowing U.S. ability will help partners. Would depend on what they are permitted to see.*
- Will it effectively support/inform Army CoE and CFT activities? *WfF and DOTMLPF-P linkages. First identifies gap. More of an identification of solution than determining capability gaps. It should be able to inform gap development.*

Ease of adoption by the training enterprise and the field:

- How readily can this be adopted by the training enterprise and the field? *Received pushback so far, but maybe just because it is new. Thinks if an approach like this was taught as another way to collect AARs* [after-action reviews] *then it would be supported but does not seem like an easy road to adopt it. It's one more thing the unit has to do, and it's looking at the problem in a more abstract way. An issue will be that the completion format is very unlike current training readiness collection methodologies.*

- How readily is this assessment integrated with DTMS? *Could be and could be valuable but has not been developed.*
- How readily is this assessment integrated with CUSR? *Interoperability is important and could be part of CUSR, but no design considerations so far.*
- How easy is it to employ during exercises? *Web access could be hard during a training event, but otherwise easy.*

Face validity of measures:

- How capable is the assessment of capturing likely limitations and possible solutions? *Relatively capable because you'll have list of limitations/solutions specifically listed in the data input (only if the observer inputs IEs). Been some discussion about having a section at the task assessment level to describe the gap/issue. We haven't put that in yet, but that could be another place where it would be required to give some description on why the assessment was given.*
- Are there known and/or reported workarounds or cheats? *Basic face validity is good, but the linkage of face validity to scale values are worrisome because there is too much subjectivity that must be used if partial scoring of level's subelements occurs.*
- What is the existing buy-in from units? From those who use the data? *Pushback so far from units, but in all the briefings for those who would use the data, everyone seems to say it would be useful (no pushback on that front)*
- Is the assessment sustainable? Statistically robust? *Could be sustainable with adequate support and resources. Statistical robustness is limited because of suspected rater agreement reliability issues.*

Likely reliability of measures and total scores:

- Is there good inter-rater reliability? *Unlikely.*
- Is there a regular review of the assessment to ensure currency and validity? *Unknown.*

Standardized and repeatable across time:

- Is the system amenable to being used across multiple exercises and years with minimal modification? *Yes.*
- Does it produce results that can be compared across the different samples? *Unlikely due to reliability issues.*

ABCANZ Armies' Program Approach

Assessment measures:

- What were the foundational concepts for developing this methodology? *ABCANZ program is operated under a governing strategy that is endorsed by an executive council. The strategy is derived from an annual program plan intended to reach program end state. The plan approves establishment of project teams whose role is to deliver measurable outputs against the Program focus areas (AACP, p. 1). Assessment is one of five steps in the End to End Mindset (AACP, p. 6).*[2]
- Describe the various elements/aspects of this methodology. *When using the PFA terminology, there are four priority focus areas. The "big four" are CIS, fires, intel fusion, and ISR. In our case, we build two of them out, so we're looking at general areas of CIS (3), IM (3), HQ (1), intel fusion (7), ISR (8), fires (5), and sustain (11). Each of these areas has a certain number of CQs noted in parentheses.*
- How many items are there to complete? *38 items to answer Y, Y(C+), or N.*
- How is the rating scale set up? What is the rating scale? (binary, categorical, text, etc.) *Y, Y(C+), or N (go into further detail below on how this works). The CQL scale is based in a single anchor point of "compatible." N means that the units were deconflicted. A Y means*

[2] *AACP* refers to Chief of Staff, American, British, Canadian, Australian, and New Zealand Armies' Program, *"Delivering an Integrated Division": The American, British, Canadian, Australian, and New Zealand Armies' Program Campaign Plan*, Arlington, Va., December 2018.

a level of compatible interoperability was achieved. A Y(C+) does not mean that integrated interoperability was achieved.

Assessment implementation:

- Who implements the assessment? *For direct collection, the ABCANZ team does, comprising about 15 members who are experienced full-time assessors. These assessments take place twice a year, although sometimes more events are budgeted in. There's also indirect collection that takes place, with reports and assessments coming in from the five nations. So, we end up with more than two assessments every year, some direct and some indirect.*
- How is the assessment implemented? *Direct collection is implemented by the ABCANZ team. Indirect collection comes from the five nations modifying their own exercises to accommodate our requests.*
- Is self-assessment permitted? *At a high level—yes.*
- What aspects of the assessment are or are not required? *For direct assessment focus events, typically, all CQL measures are collected by the ABCANZ team. That being said, it's a tailored approach, so we don't exclusively stick to CQL—we tailor each assessment to fit what's relevant and to fit within that exercise construct. For indirect focus events, ABCANZ simply leverages data collected from other observers.*
- How often do assessments occur? Are they regular, synchronized with deployment schedules, only conducted when directed? *Assessment occurs at specified training events. Those events are listed in Annex C of the AACP. The lists bifurcate exercises by direct and indirect assessment focus. Direct focus events have an ABCANZ team at them. Indirect do not have the full team. Also, for indirect events, ABCANZ relies on nondirected observations from lessons learned analysts at the exercise and so not all CQL measures may be collected.*
- How much does implementation cost? *ABCANZ relies on their own team for direct events as opposed to on-the-ground OC/T or other observers. So, the cost associated with travel, training, etc., of the ABCANZ team.*
- How easy is implementation? *Collection of the tool data is relatively straightforward. The teams also conduct internal "hotwashes" during*

exercises to achieve concurrence across tool measures and identify gaps and possible solutions.

Collection methods:

- Who collects the data/information? *The team of assessors and multiple SMEs from the countries involved for direct training events.*
- How are the data/information collected? *"Direct collect"—CQLs are a reference for that document. The indirect collection is less formal, not a singular type of information.*
- How standardized are collection measures? *Direct is more standardized than the indirect. The bulk of FQSs remain standard, but they also evolve over exercises and time.*
- At which echelon are data collected? *Division and brigade—not yet addressed for smaller units.*
- What is the cost of collection? *The cost may be expensive if applied to enduring systems with multiple raters. It really depends on the raters you choose.*
- How easy is collection, data/information capture? *Collection is easy enough. It's not challenging for the program. Unsure how much time is involved on observers' behalf to complete forms for direct collection, though claims that collecting is relatively straightforward.*
- What is the collection modality? (Subjective/objective, quantitative/qualitative) *A combination—some questions are more objective, some are more subjective. The Y/Y+/N output is quantitative while the narrative explaining the N is qualitative.*

Type of data captured:

- Are dimensions of readiness measured/captured? If so, explain. *Not a readiness measurement system. The intent of the CQL by PFA and the related FQS is to determine if "ABCANZ Armies' are becoming more interoperable" (AACP, p. 8) and is not measuring readiness.*
- Are the data captured numeric, verbal, or some combination? *Combination.*

Type of output:

- Is the output quantitative, qualitative, or some combination? *Combination.*
- What framework is the output produced within? (For example, output within AR 34-1 framework) *The strategic effort as defined by the Executive Council. This is represented by the six Priority Focus Areas. The IOC [initial operating capability] goal is to achieve a compatible level of interoperability by 2020 Q4.*
- What is the degree of standardized output? (Are results from this assessment directly comparable with outputs from other assessments?) *The output is essentially a written report with conclusions arrived at by SMEs and a judgment panel. Not directly comparable with other assessments.*

Data processing:

- To what degree is data processing automated? *No automation at all.*
- Who controls the data processing? *ABCANZ program office.*
- What types of data linkages (ABCANZ LOE, WfF, PFAs, etc.) currently exist? *ABCANZ LOE, PFAs.*

Analytics:

- To what degree is the analytics step automated? *No automation at all.*
- Who controls the analytics step? *SME observers at the event and ABCANZ program office. After initial analyses are completed, ABCANZ members may perform deep dives of the results to identify gaps and solutions.*
- How flexible is the system to make changes? *Flexible. At direct events, the ABCANZ team would be able to adjust data collection during their daily team huddles. FQSs are also updated annually, adding to flexibility. Every two years, we consider adjusting CQLs, though they're much more enduring.*

Support of strategic assessments of interoperability:

- Which parts of the road map are supported? *ABCANZ assessment was not designed or intended to directly support the road map. It is designed to support the strategy laid out by the Executive Council. This includes "the alignment of ends, ways and means balanced by risk." (AACP, p.2) The assessment will help the ABCANZ AAP identify the accomplishment of its ends through Capability and Support Groups whose activity is aimed at achieving LOE end states (AACP, p. 2). Assessment would support the ABCANZ "ways," which include leverage exercises, share best practices and lessons learned, inform doctrine, concepts and capability development, etc.*
- How close is the relationship between a road map component and assessment outputs? *ABCANZ should be map able to key elements of the road map, but how is not clear to us at this point.*
- Are measures directly/indirectly connected to DOTMLPF-P? *No—connected to PFAs. However, the results that follow training exercises are noted with possible solutions and further post-analyses analysis conducted by ABCANZ members could further define or specify DOTMLPF-P relationships.*
- Are MOPs/MOEs nested within the assessment? If not, is there a direct/indirect way to map the assessment to MOPs/MOEs? *Yes. CQLs and FQSs are nested into PFAs and the PFAs could be linked to ABCANZ LOEs.*
- Does output directly correlate to interoperability level? Or is an intermediate step needed to get from assessment output to interoperability level? *Yes. But only will identify if it is deconflicted or compatible.*

Support of other entities and partners:

- How readily does it, or could it, support senior decisionmakers? *AACP sets the strategic context, specifies the delivery framework, and details the metrics required to measure development. The results derived from AACP's measures would enable those using the method to tell which PFAs they are nearing compatibility, [and] support resourcing decisions regarding identified gaps. Assessment comes after concept and enables it to identify the nature, impact, and potential resolution of capability gaps.*

- How readily does it support our partners? *It supports Five Eyes—beyond that? It's not designed for that. We operate in the FVEY* [Five Eyes] *community. That being said, this doesn't prohibit the methodology and questions from being used. In principle, this could be utilized beyond the FVEY.*
- Will it effectively support/inform Army CoE and CFT activities? *It can inform them through HQDA in that the United States is an ABCANZ partner and may identify a desire to conduct a deep dive on an identified capability gap that is found in the assessment process.*

Ease of adoption by the training enterprise and the field:

- How readily can this be adopted by the training enterprise and the field? *Unlikely to be adopted by U.S. forces without significant changes to training process. However, ABCANZ can run parallel to United States as well as other partners' training enterprise activities.*
- How readily is this assessment integrated with DTMS? *Would not be.*
- How readily is this assessment integrated with CUSR? *Would not be.*
- How easy is it to employ during exercises? *Easy, as long as ABCANZ team is present.*

Face validity of measures:

- How capable is the assessment of capturing likely limitations and possible solutions? *Very.*
- Are there known and/or reported workarounds or cheats? *The team huddles provide flexibility and a means of achieving agreement among raters/observers. This should help bolster reliability.*
- What is the existing buy-in from units? From those who use the data? *Because the collection team is on the ground doing it themselves, buy-in doesn't have to be as extensive. People also aren't overly nervous about getting a bad mark when it comes to interoperability since it's seen as a "non-testing non-evaluating area." Where buy-in does become an issue is in the effort and emphasis Chief of Staff or command group applies*

to embracing these standards. This can affect the quality of results. We see that smaller nations are more capable of plugging in ABCANZ standards and working within the ABCANZ construct than, perhaps, a U.S. division or brigade.

- Is the assessment sustainable? Statistically robust? *Sustainable as long as it is resourced. Statistically as robust as other systems. The observer huddles should improve rater agreement.*

Likely reliability of measures and total scores:

- Is there good inter-rater reliability? *Yes. Decisions are agreed upon by the group in the FVEY. We have multiple judges come together, which is how we overcome subjectivity.*
- Is there a regular review of the assessment to ensure currency and validity? *Yes.*

Standardized and repeatable across time:

- Is the system amenable to being used across multiple exercises and years with minimal modification? *Yes.*
- Does it produce results that can be compared across the different samples? *Technically, yes, and the intent to compare is there, but it's not as easy to compare as other assessments may be. You just need to understand the context of each exercise and report, as opposed to simply directly comparing data.*

Multinational Interoperability Assessment Tool

Assessment measures:

- What were the foundational concepts for developing this methodology? *Put science into the art of determining interoperability.*
- Describe the various elements/aspects of this methodology. *Using ABCANZ CQL, provide yes/no answers. Yes/no responses are linked to levels of interoperability (e.g., if you get yes on 1–4, you're deconflicted; if you get yes on 5–9, you're compatible; 9–12, you're integrated). At the*

time of this data collection, "integrated" was not used; the study team was told that this has since changed as of late 2020 to include "integrated."

- How many items are there to complete? *Variable, depending on operation and/or phase within operation.*
- How is the rating scale set up? What is the rating scale? (binary, categorical, text, etc.) *Binary questionnaire which is linked to ABCANZ interoperability levels.*

Assessment implementation:

- Who implements the assessment? *Tool will be simple enough that anyone (regardless of country, echelon, position, etc.) can pick it up and fill it in. Can be self-assessment or bilateral/multinational assessment. Goal is that it will be intuitive enough that you don't need to train raters.*
- How is the assessment implemented? *Assessments filled out on laptops, results go to local laptop database, analysis done separately.*
- Is self-assessment permitted? *Yes—in fact, it is encouraged.*
- What aspects of the assessment are or are not required? *Different reports/items filled out based on the exercise or phase/activity during exercise.*
- How often do assessments occur? Are they regular, synchronized with deployment schedules, only conducted when directed? *At any exercise with MNI component, throughout the exercise.*
- How easy is implementation? *Goal is for it to be easy and intuitive.*

Collection methods:

- Who collects the data/information? *Anyone.*
- How are the data/information collected? *Collected on laptop, then downloaded to central location for analysis.*
- How standardized are collection measures? *Completely standardized.*
- At which echelon are data collected? *Designed for higher levels (BDE* [brigade] *and above) but will have stretch-downwards potential.*
- What is the collection modality? (Subjective/objective, quantitative/qualitative) *Objective, quantitative where possible (e.g., fires delivered within certain amount of time).*

Type of data captured:

- Are dimensions of readiness measured/captured? If so, explain. *Interoperability is component of readiness but will not report on readiness per se.*
- Are the data captured numeric, verbal, or some combination? *Mostly numeric or binary (yes/no) but with optional narrative portions.*

Type of output:

- Is the output quantitative, qualitative, or some combination? *Objective, quantitative where possible (e.g., fires delivered within certain amount of time).*
- What framework is the output produced within? (For example, output within AR 34-1 framework) *ABCANZ campaign plan interoperability levels.*
- What is the degree of standardized output? (Are results from this assessment directly comparable with outputs from other assessments?) *Highly standardized.*

Data processing:

- To what degree is data processing automated? *Headline results will be automated.*
- Who controls the data processing? *Contractor (to be determined) will control analytics, but Army will own/control data and results.*
- What types of data linkages (ABCANZ LOE, WfF, PFAs, etc.) currently exist? *ABCANZ LOE, others could make sense.*

Analytics:

- To what degree is the analytics step automated? *Headline results will be automated.*
- Who controls the analytics step? *Contractor (to be determined) will control analytics, but Army will own/control data and results.*

- How flexible is the system to make changes? *Very flexible. Army owns intellectual property rights, so do not need to go through contractor to make changes.*

Support of strategic assessments of interoperability:

- How are strategic goals supported? *It will measure where we are and see if they line up to visions of end-state.*

Support of other entities and partners:

- How readily does it, or could it, support senior decisionmakers? *Enable senior leaders to make balanced investment decisions (or decide to carry the risk). Also just to report back.*
- Will it effectively support/inform Army CoE and CFT activities? *UK equivalent of these organizations do equipment planning, balance of investments, etc., and we expect this to inform those decisions. Also contributes on procedural side and technical side. MIAT will say, "this is where you're at," and you'll have to deduce where you're not at.*

Ease of adoption by the training enterprise and the field:

- How readily can this be adopted by the training enterprise and the field? *No vision for it to be used in training but could be used in equipment trials.*
- How readily is this assessment integrated with DTMS? *UK equivalent: could be useful, but isn't a priority.*
- How readily is this assessment integrated with CUSR? *It's a component of readiness, but we're not measuring readiness.*
- How easy is it to employ during exercises? *Very easy.*

Face validity of measures:

- How capable is the assessment of capturing likely limitations and possible solutions? *Not set up to do this because raters are not necessarily experts in these areas.*

- Are there known and/or reported workarounds or cheats? *Not yet implemented, so unsure.*
- What is the existing buy-in from units? From those who use the data? *System has not been shared with anyone yet, so unsure.*
- Is the assessment sustainable? Statistically robust? *Goal is for it to be sustainable and statistically robust.*

Likely reliability of measures and total scores:

- Is there good inter-rater reliability? *No plans to assess this.*
- Is there a regular review of the assessment to ensure currency and validity? *Before fielding, will have to be frequent review. After fielding, annual review will probably be sufficient.*

Standardized and repeatable across time:

- Is the system amenable to being used across multiple exercises and years with minimal modification? *Yes.*
- Does it produce results that can be compared across the different samples? *Yes.*

APPENDIX C

ART Level I and Level II Tasks Included in AIMS Instruments

This appendix shows the Army Level I and Level II tasks, and whether they were included in both Pilots 1 and 2.

TABLE C.1
ART Tasks

ART Level I	ART Level II	Pilot 1	Pilot 2
1.0 Movement and Maneuver	1.1: Perform tactical actions associated with force projection and deployment		
	1.2: Conduct tactical maneuver		X
	1.3: Conduct tactical troop movements		X
	1.4: Conduct direct fires		X
	1.5: Occupy an area		X
	1.6: Conduct mobility operations		X
	1.7: Conduct countermobility operations		X
	1.8: Conduct reconnaissance		X
	1.9: Employ obscurants		X
	1.10: Conduct maneuver support operations		X
2.0 Intelligence	2.1: Provide intelligence support to force generation	X	X
	2.2: Provide support to situational understanding	X	X
	2.3: Conduct information collection	X	X
	2.4: Provide intelligence support to targeting and information-related capabilities	X	X

Table C.1—Continued

ART Level I	ART Level II	Pilot 1	Pilot 2
3.0 Fires	3.1: Integrate fires	X	X
	3.2: Provide fire support	X	X
	3.3: Integrate air-ground operations	X	X
	3.4: Employ air and missile defense	X	X
4.0 Sustainment	4.1: Provide logistics support	X	X
	4.2: Provide personnel support	X	X
	4.3: Provide health service support	X	X
	4.4: Provide financial management support		
5.0 Mission Command	5.1: Conduct the operations process	X	X
	5.2: Conduct command post operations	X	X
	5.3: Conduct knowledge management and information management	X	X
	5.4: Control tactical airspace		
	5.5: Execute command programs		
	5.6: Integrate space operations		
	5.7: Conduct public affairs operations		
	5.8: Develop teams		
	5.9: Conduct cyber electromagnetic activities		
	5.10: Install, operate, and maintain the network	X	X
	5.11: Conduct military deception		X
	5.12: Synchronize information-related capabilities		
	5.13: Conduct soldier and leader engagement		
	5.14: Employ military information support operations		
	5.15: Conduct civil affairs operations		X

Table C.1—Continued

ART Level I	ART Level II	Pilot 1	Pilot 2
6.0 Protection	6.1: Coordinate air and missile defense	X	X
	6.2: Conduct personnel recovery operations	X	X
	6.3: Implement physical security procedures		
	6.4: Conduct operational area security		
	6.5: Apply antiterrorism measures		X
	6.6: Conduct survivability operations	X	X
	6.7: Provide force health protection		X
	6.8: Conduct chemical, biological, radiological	X	X
	6.9: Employ safety techniques	X	X
	6.10: Implement operations security		
	6.11: Provide explosive ordnance disposal		X
	6.12: Conduct detention operations		
	6.13: Conduct police operations		X
	6.14: Conduct resettlement operations		

SOURCE: Provided by HQDA, April 23, 2019.

APPENDIX D

Interoperability in Army Mission Essential Tasks

Recently, the Army explored options for including MNI in unit METLs at echelons above brigade. As a result, corps- and theater Army (TA)–level METs were revised to report specifically against MNI aspects when units conduct MNI exercises.[1] The following verbiage (in **bold**) was added to the conditions statement in relevant METs and SCTs, with appropriate modifications based on whether it was written for corps, TA, or joint forces:

> The [corps, theater army (TA), joint force] may execute some iterations of this task with a multinational (MN) component to the force. **When the [corps, TA, joint force] is training this task with an MN partner, evaluate all MNI-related steps [and] measures in this T&EO. For the purpose of this requirement, the MN partner is a force of brigade or larger size that reports directly to [the corps, the TA, a component of the joint force] and has interoperability focus area capabilities (fires, intelligence, sustainment, mission command).** When the [corps, TA, joint force] is executing this task in a scenario without a multinational component, evaluators should rate steps in this task that only apply to multinational operations scenarios as N/A.

An interoperability critical step or measure was also added to relevant METs and SCTs in the "Prepare" section. Figure D.1 is an example of the

[1] HQDA G-3/5/7, "Incorporating Multinational Interoperability Conditions into Corps and Theater Army METs," DAMO-TR Brief, August 17, 2018, shared with authors by MCCoE on March 1, 2019, Not available to the general public.

interoperability measure as added to MET 71-CORPS-1270: Conduct Shaping Operations for Corps.[2]

Table D.1 lists the corps, TA METs, and SCTs that include MNI considerations. MNI considerations are not present in METs related to deployment/force-projection tasks, noncombatant evacuation operations tasks, or defense support of civil authorities tasks.

FIGURE D.1

Example Excerpt of Interoperability Critical Step/Measure

+ 4. The corps staff, led by the chief of staff, establishes interoperability with multinational forces, including:

a. The G-6 implements the directed multinational mission command information system(s) established for the operation in accordance with established policy and guidance.

b. The G-2 establishes intelligence fusion procedures including multinational capabilities.

c. The knowledge management officer and G-2 establish information sharing procedures for multinational partners.

d. The G-2 incorporates multinational partner reconnaissance and surveillance capabilities into the intelligence collection plan.

e. The G-3 incorporates multinational forces into the common operating picture (COP) and other current operations monitoring, reporting, and analysis procedures.

f. The chief of fires establishes a system (people, mission command information system, facilities, processes, procedures) for planning and executing joint fires.

g. The G-4 synchronizes logistics with host nation (HN) and multinational (MN) partners.

h. The chief of protection coordinates security and protection with HN and MN partners.

i. The knowledge management officer (KMO) incorporates multinational information requirements into the knowledge management plan.

j. The foreign disclosure officer and KMO incorporates foreign disclosure procedures into the knowledge management plan.

k. The G-2 coordinates linguist and translator support.

l. The COS establishes liaison with multinational forces.

Note: Evaluate this step only when the corps is part of a multinational force. See collective task 71-CORP-5725, conduct multinational operations, for additional requirements for interoperability.

m. The command maintains operations security over information and operations that are not releasable to foreign nationals (NOFORN).

SOURCE: U.S. Army, MET 71-CORPS-1270: Conduct Shaping Operations for Corps (see Fowler, 2021).

[2] Mission Essential Tasks can be accessed by approved people through the Army Training Network at atn.army.mil. See Andrew Fowler, "The Army Training Network: Gateway to Training Management," U.S. Army webpage, January 12, 2021.

TABLE D.1

Mission-Essential Tasks and Supporting Collective Tasks with Multinational Interoperability Considerations

MET or SCT	Number	Title	ARCENT CCP	ARNORTH CCP	ARSOUTH CCP	ASCC MCP and 8th Army	Corps	USAREUR CCP and MCP	USARPAC
MET	71-CMD-5100	Conduct the Mission Command Operations Process for Commands	X						
MET	71-CORP-1270	Conduct Shaping Operations for Corps					X		
MET	71-CORP-1340	Conduct Forcible Entry Operations for Corps					X		
SCT	71-CORP-5100	Conduct the Mission Command Operations Process for Corps					X		
MET	71-CORP-7000	Conduct Campaign and Major Land Combat Operations for Corps					X		
MET	71-CORP-7120	Conduct an Attack for Corps					X		
MET	71-CORP-7222	Conduct a Defense for Corps					X		

Table D.1—Continued

MET or SCT	Number	Title	ARCENT CCP	ARNORTH CCP	ARSOUTH CCP	ASCC MCP and 8th Army	Corps	USAREUR CCP and MCP	USARPAC
MET	71-JNT-2700	Provide Intelligence Support to Joint Operations						X	
MET	71-JNT-3100	Conduct Joint Force Targeting for Joint Task Force	X						
MET	71-JNT-3300	Conduct Peace Operations for Joint Task Force						X	
MET	71-JNT-4001	Coordinate Joint Logistics for Joint Task Force						X	
MET	71-JNT-5100	Conduct Joint Operations Processes for Joint Task Force	X	X	X			X	X
MET	71-JNT-5500	Establish a Joint Task Force Headquarters	X						
MET	71-JNT-6000	Conduct Operational Force Protection for Joint Task Force						X	

Table D.1—Continued

MET or SCT	Number	Title	ARCENT CCP	ARNORTH CCP	ARSOUTH CCP	ASCC MCP and 8th Army	Corps	USAREUR CCP and MCP	USARPAC
MET	71-JNT-6200	Provide Protection for Operational Force, Means, and Noncombatants for the Joint Force						X	
MET	71-TA-1130	Conduct Joint Reception, Staging, Onward Movement and Integration for Theater Army				X			
MET	71-TA-2500	Establish Intelligence Enterprise Interoperability for Theater-Level Operations for Theater Army				X			
MET	71-TA-5100	Conduct the Mission Command Operations Process for Theater Army				X			
MET	71-TA-5450	Coordinate Support for Forces in Theater for Theater Army				X		X	

Table D.1—Continued

MET or SCT	Number	Title	ARCENT CCP	ARNORTH CCP	ARSOUTH CCP	ASCC MCP and 8th Army	Corps	USAREUR CCP and MCP	USARPAC
MET	71-TA-5711	Conduct Theater Security Cooperation for Theater Army			X	X		X	
MET	71-TA-7320	Conduct Limited Contingency Operations for Theater Army Contingency Command Post	X		X			X	X

SOURCE: Documents provided by the Army.
NOTES: ARCENT = U.S. Army Central; ARNORTH = U.S. Army North; ARSOUTH = U.S. Army South; ASCC MCP = Army Service Component Command Main Command Post; CCP = Contingency Command Post; MCP = Main Command Post; USAREUR = U.S. Army Europe and Africa; USARPAC = U.S. Army Pacific.

APPENDIX E

Computing Priority Force Area Interoperability Levels

This appendix provides the current method of aggregating individual measures into a singular PFA level value. Each measure is associated with a single interoperability level value or one of several possible level values. For measures with a yes or no dichotomous option, the possible value is a singular level of 1, 2, or 3, depending on the CoEs' determination based on the original interoperability level for the respective ART Level II task. Measures with a select-one-response stem-and-root response format have several possible level values based on the original determination from the CoEs. For instance, if underneath a stem there are four options ("roots"), then there would be four levels possible. The level for each root is based on the interoperability ART Level II task definitions.

For each measure on a PFA instrument, the recorded response is compared with the level associated with the measure. For dichotomous measures, if the response is yes, then that measure receives the value of the level associated with the measure. For select-one-response measures, the recorded response receives the level associated with that response and all other nonzero level values lower than the selected value. In other words, for a select one–style measure, if a 3 is the recorded level, a 3, 2, and 1 would be recorded. For an example of this selection step, refer to Table E.1 (a duplicate of Table 5.3). In Table E.1, measure 1 is a select-one format option. If b were the recorded option, then the values of 2 and 1 would be included in the PFA level calculation. Measure 4 is a dichotomous measure; if a yes were recorded, a 3 would be the value, and if a no were recorded, a 0 would be the value for that measure. Measures recorded N/A are excluded from the PFA-level calculations.

TABLE E.1
Stem/Root System of ART 3.2

ART 3.2 Interoperability Measures	I-Level
1. Select the one that best describes Army and MN partners' network in terms of employing fire support (select one).	
(a) Use organic digital capabilities that facilitate the collective and coordinated use of indirect, joint, and electronic fires against surface targets to include nonlethal capabilities.	3
(b) Leverages a combination of digital and analog capabilities or methodologies. Networked capabilities are limited to select digital-to-digital touch points, often driving human intervention to conduct digital-to-analog-to-digital conversions with some LNO participation.	2
(c) Leverages analog systems or leverages with outdated, cyber-vulnerable digital capability.	1
(d) Does not integrate fires using any digital, analog, or liaison means of communications.	0
2. Select the one that best describes the conduct of D3A targeting process U.S. and MN partners (select one).	
(a) Send digital information in near real time that supports D3A using each mission partner's organic fires C2 systems and support structure.	3
(b) Employ the MPE but rely on some human dimensions to incorporate procedural controls IOT support D3A.	2
(c) Require analog-to-digital conversions to support D3A.	1
(d) U.S. and MN partners did not integrate fires.	0
3. Select the one that best describes LNO involvement during fires integration (select one).	
(a) U.S. and MN partners successfully integrate fires with little to no LNO support necessary.	3
(b) Liaison with U.S. network equipment required.	2
(c) Rely heavily on liaison teams resourced with U.S. equipment and personnel.	1
(d) Did not establish liaisons: they operate independently.	0
4. U.S. and MN partners nominate and achieve electronic and computer network attack effects on targets.	3

SOURCE: RAND Arroyo Center modification of HQDA content.

The first step is to determine the number of ***possible*** 1s, 2s, and 3s by counting the number of times each of those values could have occurred (again, if a measure is recorded as N/A, it is not included in the count) across all measures for a PFA. The following list is an example of counting the possible levels:

- CIS PFA has 20 measures.
- For an exercise, two measures were recorded as N/A because the exercise did not stimulate that activity.
- There are 18 possible measures.
  - of these 18 measures, ten are select-one-option measures each with a value of 0–3 being possible, so there are ten 3s, 2s, and 1s possible.
  - The other eight measures are dichotomous; of these, five measures have a determined level equal to 3 and the other three have a level equal to 2.
  - In total, for the 18 measures, there are fifteen 3s, thirteen 2s, and ten 1s possible.

Next, count the number of ***received*** 1s, 2s, and 3s that are recorded. Continuing the example above:

- There were 18 measures recorded.
- For the ten select-one measures, there were
  - five that were recorded as a 3. For these five measures, you count in the number that received a 1, 2 and 3.
  - five that were recorded a 2. For these measures, you count in the number that received a 1 and 2.

- For the five dichotomous measures with a possible 3, there were
  - two recorded yes, so two 3s.
  - three recorded no, so these are not counted as received.

- For the three dichotomous measures with a possible 2, all were recorded yes, so there are three more 2s received.
- In total, across all 18 measures, seven 3s, twelve 2s, and ten 1s were received.

Now, compute the quotient for the 1s, 2s, and 3s by dividing the count of ***received*** levels by the count of ***possible*** levels. Continuing with the example, there were

- fifteen 3s possible and seven received, so 7/15 = 0.46
- thirteen 2s possible and 12 received, so 12/13 = 0.92
- ten 1s possible and ten received, so 10/10 = 1.00.

To determine the PFA interoperability level value score, apply the following rules (where *Q* equals quotient):

- If Q1 < 0.5, the level is 0.
- If 0.50 < Q1 < 0.70, the level is 1(-).
- If Q1 > 0.70 AND Q2 < 1.00, sum Q1 and Q2.
  - If 1.00 < (sum Q1 Q2) < 1.50, the level is 1.
  - If 1.50 < (sum Q1 Q2) < 1.99, the level is 1(+).

- If Q1 > 0.70 AND Q2 = 1.00, sum Q1, Q2, and Q3.
  - If (sum of Q1, Q2, Q3) < 2.00, the level is 2(-).
  - If 2.00 < (sum of Q1, Q2, Q3) < 2.25, the level is 2.
  - If 2.25 < (sum of Q1, Q2, Q3) < 2.75, the level is 2(+).
  - If 2.75 < (sum of Q1, Q2, Q3) < 2.99, the level is 3(-).
  - If (sum of Q1, Q2, Q3) = 3.00, the level is 3.

Abbreviations

7ATC	7th Army Training Command
AACP	ABCANZ Armies' Program Campaign Plan
AAP	ABCANZ Armies' Program
ABCANZ	American, British, Canadian, Australian, and New Zealand
AIMS	Army Interoperability Measurement System
AoA	analysis of alternatives
AR	Army Regulation
ART	Army tactical task
BCT	brigade combat team
C2	command and control
CAA	Center for Army Analysis
CAAT	collection and analysis team
CALL	Center for Army Lessons Learned
CFT	cross-functional team
CIRCuIT	Communications Interoperability Capability Appraisal Table
CIS	communication and information systems
CoE	center of excellence
CQL	critical question list
CUSR	commander unit status report
D3A	decide, detect, deliver, assess

DAMO-SSC	Department of the Army, Military Operations—Stability and Security Cooperation Division
DOTMLPF-P	doctrine, organization, training, materiel, leadership and education, personnel, facilities, and policy
DTMS	digital training management system
EXORD	execute order
FQS	focused question set
FVEY	Five Eyes; an intelligence partnership comprising Australia, Canada, New Zealand, the United Kingdom, and the United States
HQDA	Headquarters Department of the Army
IM/KM	information management and knowledge management
ISR	intelligence, surveillance, and reconnaissance
JMC	Joint Modernization Command
JMRC	Joint Multinational Readiness Center
JWA	Joint Warfighting Assessment
LNO	liaison officer
LOE	line of effort
MCCoE	Mission Command Center of Excellence
MET	mission-essential task
METL	mission-essential task list
MFC	multinational fusion cell
MIAT	Multinational Interoperability Assessment Tool
MN	multinational
MNI	multinational interoperability

MOE	measure of effectiveness
MOP	measure of performance
MPE	mission partner environment
NDS	National Defense Strategy
OC/T	observer, coach, trainer
PFA	priority focus area
ROD	record of decision
SCT	supporting collective task
SME	subject-matter expert
T&EO	training and evaluation outline
TA	theater Army
TRADOC	U.S. Army Training and Doctrine Command
TS 19	Talisman Sabre 19
UAP	unified action partner
WfF	warfighting function

References

AR—*See* Army Regulation.

Army Regulation 11-33, *Army Lessons Learned Program*, Washington, D.C.: Headquarters, Department of the Army, June 2018.

Army Regulation 34-1, *Interoperability*, Washington, D.C.: Headquarters, Department of the Army, April 9, 2020.

Army Regulation 220-1, *Army Unit Status Reporting and Force Registration—Consolidated Policies,* Washington, D.C.: Headquarters, Department of the Army, April 15, 2010. As of April 4, 2021:
https://armypubs.army.mil/epubs/DR_pubs/DR_a/pdf/web/r220_1.pdf

Chief of Staff, American, British, Canadian, Australian, and New Zealand Armies' Program, *"Delivering an Integrated Division": The American, British, Canadian, Australian, and New Zealand Armies' Program Campaign Plan*, Arlington, Va., December 2018.

EXORD—*See* U.S. Department of the Army, Execute Order.

Fowler, Andrew, "The Army Training Network: Gateway to Training Management," U.S. Army webpage, January 12, 2021. As of April 15, 2021:
https://www.army.mil/article/242347/
the_army_training_network_gateway_to_training_management

Headquarters, Department of the Army, G-3/5/7, "Incorporating Multinational Interoperability Conditions into Corps and Theater Army METs," DAMO-TR Brief, August 17, 2018, shared with authors by MCCoE on March 1, 2019, Not available to the general public.

Headquarters, Deputy Chief of Staff, G-3/5/7, "The Army Strategy," U.S. Army webpage, October 25, 2018. As of March 31, 2021:
https://www.army.mil/standto/archive/2018/10/25/

Keys, Ronald, John Castellaw, Robert Parker, Jonathan White, Gerald Galloway, and Christine Parthemore, *Military Expert Panel Report: Sea Level Rise and the U.S. Military's Mission*, Shiloh Fetzek, Caitlin E. Werrell, and Francesco Femia, eds., Washington, D.C.: Center for Climate and Security, September 2016. As of March 31, 2021:
https://climateandsecurity.files.wordpress.com/2016/09/
center-for-climate-and-security_military-expert-panel-report_brochure.pdf

Thomas, Jon T., and Douglas L. Schultz, "Lessons About Lessons: Growing the Joint Lessons Learned Program," *Joint Force Quarterly*, Vol. 79, 4th Quarter, October 1, 2015, pp. 113–120. As of March 31, 2021:
https://ndupress.ndu.edu/JFQ/Joint-Force-Quarterly-79/Article/621147
/lessons-about-lessons-growing-the-joint-lessons-learned-program/

U.S. Army Combined Arms Center, *Conduct Joint Operations Processes for Joint Task Force*, 71-JNT-5100, Fort Leavenworth, Kan., October 2019.

U.S. Army Combined Arms Center, Center for Army Lessons Learned, *Multinational Interoperability: Reference Guide—Lessons and Best Practices*, Handbook No. 16-18, Fort Leavenworth, Kan., July 2016. As of September 17, 2019:
https://usacac.army.mil/sites/default/files/publications/16-18.pdf

U.S. Army Training and Doctrine Command, TRADOC Regulation 25-36, *The TRADOC Doctrine Publication Program*, Fort Eustis, Va., May 21, 2014.

U.S. Department of the Army, *Leader's Guide to Objective Assessment of Training Proficiency*, Washington, D.C., September 2017.

U.S. Department of the Army, Execute Order 293-17, *Multinational Interoperability Training*, Washington, D.C., September 2017.

U.S. Department of the Army, G-3/5/7, *DAMO-TR Briefing: Incorporating Multinational Interoperability Conditions into Corps and Theater Army METs*, Washington, D.C., August 2018, Not available to the general public.

U.S. Department of Defense, *Summary of the National Defense Strategy of the United States of America*, Washington, D.C., 2018. As of September 17, 2019:
https://dod.defense.gov/Portals/1/Documents/pubs/2018-National-Defense-Strategy-Summary.pdf